Sustaining Tomorrow via Innovative Engineering

Highly Recommended Titles

Water Is ... : The Indispensability of Water in Society and Life
by Seth B Darling, Seth W Snyder
ISBN: 978-981-3271-39-5
ISBN: 978-981-3278-10-3 (pbk)

Science of the Earth, Climate and Energy
by Milton W Cole, Angela D Lueking and David L Goodstein
ISBN: 978-981-3233-61-4
ISBN: 978-981-3271-78-4 (pbk)

The Energy Conundrum: Climate Change, Global Prosperity, and the Tough Decisions We Have to Make
by Neil A C Hirst
ISBN: 978-1-78634-460-1
ISBN: 978-1-78634-667-4 (pbk)

The Earth as a Cradle for Life: The Origin, Evolution and Future of the Environment
by Frank D Stacey and Jane H Hodgkinson
ISBN: 978-981-4508-32-2

The Goldilocks Policy: The Basis for a Grand Energy Bargain
by John R Fanchi
ISBN: 978-981-3276-39-0
ISBN: 978-981-3277-44-1 (pbk)

Saving Ourselves: Interviews with World Leaders on the Sustainable Transition
edited by Yacine Belhaj-Bouabdallah
ISBN: 978-981-3220-74-4
ISBN: 978-981-3220-75-1 (pbk)

Sustaining Tomorrow via Innovative Engineering

Editors

David S-K Ting
University of Windsor, Canada

Rupp Carriveau
University of Windsor, Canada

World Scientific

NEW JERSEY · LONDON · SINGAPORE · BEIJING · SHANGHAI · HONG KONG · TAIPEI · CHENNAI · TOKYO

Published by

World Scientific Publishing Co. Pte. Ltd.

5 Toh Tuck Link, Singapore 596224

USA office: 27 Warren Street, Suite 401-402, Hackensack, NJ 07601

UK office: 57 Shelton Street, Covent Garden, London WC2H 9HE

Library of Congress Cataloging-in-Publication Data

Names: Ting, David S-K, editor. | Carriveau, Rupp, editor.
Title: Sustaining tomorrow via innovative engineering / editors, David S-K Ting,
    University of Windsor, Canada, Rupp Carriveau, University of Windsor, Canada.
Description: Singapore ; Hackensack, NJ ; London : World Scientific, [2021] |
    Includes bibliographical references.
Identifiers: LCCN 2021058776 | ISBN 9789811228025 (hardcover) |
    ISBN 9789811228032 (ebook for institutions) | ISBN 9789811228049 (ebook for individuals)
Subjects: LCSH: Energy conservation. | Sustainable engineering. | Renewable energy sources.
Classification: LCC TJ163.3 .S896 2021 | DDC 338.9/27--dc23
LC record available at https://lccn.loc.gov/2021058776

British Library Cataloguing-in-Publication Data
A catalogue record for this book is available from the British Library.

For any available supplementary material, please visit
https://www.worldscientific.com/worldscibooks/10.1142/12033#t=suppl

Desk Editors: Britta Ramaraj/Amanda Yun

Typeset by Diacritech Technologies Pvt. Ltd.
Chennai - 600106, India

To every soul who strives to
sustain tomorrow

Preface

Abraham Lincoln asserted, "You cannot escape the responsibility of tomorrow by evading it today." Rather, we gladly follow Elizabeth Barrett Browning and "Light tomorrow with today." The question then is how do we light tomorrow with today? Are renewables the only way to light tomorrow's energy? What about less privileged nations? The answers to these questions are contained within Reader's straightforward question, "Energy: A Reasonable Mix?" in Chapter 1. If you are still unconvinced that fossil fuels are inevitable in furnishing a tomorrow for a significant portion of the world's population, allow Tchanche to enlighten you with plain numbers concerning "Energy Supply and Consumption in Senegal" in Chapter 2. Senegal is unique as a country but it also serves as an example of a developing nation endeavoring to improve the living standard by providing energy to its populace. In Chapter 3, "Hydrogen as an Energy Vector: Present and Future," Llamas et al. reveal that currently about 95% of hydrogen produced for energy is from fossil fuels. Hydrogen's potential lies in its capacity as an energy storage medium to further enable the engagement of intermittent renewable energy generation. The easiest way to tap into free solar energy is to harness its thermal energy for heating applications. Lanjewar and Patel detail how to enhance solar air heater efficiency in Chapter 4, "V-Shaped Roughened Geometries and Their Effect on Heat Transfer and Friction Factor in Solar Air Heater." The idea is to employ V-rib geometries to boost heat convection while minimizing the inevitable friction increase. While discussion of leading edge energy efficiency in new construction is very popular, the fact remains that many buildings last for decades. Thus, appropriate retrofitting of aging buildings must be a part of the equation in sustaining tomorrow. This is particularly true for tall and supertall buildings in big cities. The challenge and the solutions associated with green retrofitting these man-made wonders are expounded by Al-Kodmany in Chapter 5, "Retrofitting Aging Tall and Supertall

Buildings: a Continuing Challenge." Discussion of buildings reminds readers that they are the very essence of tradition and culture for many peoples. Ergin et al. illuminate ways of creating livable and energy-efficient space in Chapter 6, "Energy Efficiency as a Function of Changing Building Form in Sustainable Rural Architecture." The key is to implement energy efficiency without removing the unique architectural identities of the tradition and culture of the particular region. There is no tomorrow to sustain if there is no clean air to breathe or no potable water to drink. Brimblecombe fittingly suggests legal and fiscal responses, in addition to technical controls, to solve the air and water pollution problem in Chapter 7, "Sustaining our Air and Water." After delving into the large scale and scope of the issue, he sheds light on dealing with the problem, i.e., sustainable engineering design, soft engineered solutions, along with economic instruments and sociological change. Other than potable water, the vast volume of saltwater is critical to both lives on land and in the water. Chou proposes living seawalls for mitigating rising sea levels by eco-engineering intertidal pools and lagoons for restoring coastal and nearshore habitats in Chapter 8, "'Living' Coastal Protection and Nearshore Biodiversity Reclamation." The book concludes with "Transitioning Toward Sustainable Development Through the Water–Energy–Food Nexus" by Nhamo et al. as Chapter 9. To highlight, zero hunger, clean water and sanitation, and affordable and clean energy are key elements of sustainable development goals that acknowledge the interlinkages between human well-being, economic prosperity, and a healthy environment. This volume alone, as willing as all involved are, cannot fulfill that big dream of sustaining tomorrow. Our hope is to create some positive ripples, even if they are much smaller than those created by Mother Teresa, who walked beyond what she uttered, "I alone cannot change the world, but I can cast a stone across the waters to create many ripples."

David S-K. Ting and Rupp Carriveau
Turbulence and Energy Laboratory
Allinterest Research Institute

Acknowledgments

This book could not be accomplished apart from abundant support, including the indispensable grace from above, lavished upon the editors. The editors are exceedingly grateful to the amiable team of World Scientific Publishing, Publishing Director, Hong Koon Chua, and Senior Editor, Amanda Yun, in particular. There would have been not much to convey without the quality contributions from the experts who penned the many chapters, and the constructive reviewers who exercised quality control in Sustaining Tomorrow via Engineering Innovation.

Table of Contents

Chapter 1.   Energy: A Reasonable Mix?.................................. 1
Graham T. Reader

Chapter 2.   Energy Supply and Consumption
in Senegal..55
Bertrand Tchanche

Chapter 3.   Hydrogen as an Energy Vector:
Present and Future..83
*Bernardo Llamas, Javier Garciat, Marcelo F. Ortega Romero,
Isabel Amez, M. Belen Vallespir Jimenez*

Chapter 4.   V-shaped Roughened Geometries and Their Effect
on Heat Transfer and Friction Factor in Solar Air
Heater 131
Atul Lanjewar and Sumer Singh Patel

Chapter 5.   Retrofitting aging Tall and Supertall Buildings:
A Continuing Challenge................................... 171
Kheir M. Al-Kodmany

Chapter 6.   Energy Efficiency as a Function of Changing
Building Form in Sustainable
Rural Architecture 217
Şefika Ergin, Kemal Çorapçioğlu, Figen Balo

Chapter 7.   Sustaining Our Air and Water................................ 259
Peter Brimblecombe

Chapter 8. "Living" Coastal Protection and Nearshore Biodiversity Reclamation 291

Loke Ming Chou

Chapter 9. Transitioning Toward Sustainable Development Through the Water-Energy-Food Nexus 311

Luxon Nhamo, Sylvester Mpandeli, Aidan Senzanje, Stanley Liphadzi, Dhesigen Naidoo, Albert T. Modi, Tafadzwanashe Mabhaudhi

CHAPTER 1
Energy: A Reasonable Mix?

Graham T. Reader

*Department of Mechanical, Automotive and Materials Engineering,
University of Windsor, Ontario, Canada*

Abstract

As the global population continues to increase, so does the demand for energy. It is unlikely that this increased demand can be met by the ubiquitous fossil fuel sources alone and, therefore, the use of other energy sources will need to expand. Fortunately, there are numerous sources of energy and many different methods for converting that energy into useful forms, like electricity. Indeed, the United Nations is seeking to increase access to electricity by all people as part of their overarching goal to eliminate global poverty. At the same time, because of the projections and observations surrounding climate change models, the elimination of fossil fuels, and their "harmful" atmospheric emissions, has been promoted. Fossil fuels would be replaced by sustainable and renewable energy sources such as solar and wind power. Some of the richer countries, like Germany and Denmark, have already acted vigorously on such replacements. However, the largest users of energy and producers of emissions still only obtain a small, but increasing, percentage of their energy from nonrenewable sources. Moreover, despite the laudable United Nations (UN) goals, and those of the "Paris Agreement" over a third of the global population cannot access clean cooking fuels, and many millions still live in abject poverty. In such circumstances is it possible to have a global agreement on a universally acceptable energy mix, which includes a complete transition away from fossil fuels? If not, what are the alternatives? These questions are explored in this chapter.

Keywords: Energy conversion, renewable energy, energy transitions

1.0   Introduction

There are many sources of energy and many different methods for converting that energy into useful forms, such as electricity. The energy sources that are used today, except for nuclear fission, have been used for over 2.5 millennia to a lesser or greater extent in the historic *energy mixes* [1]. Until the 18th century's industrial revolution, the dominant energy source was biomass (wood), today, other carbonaceous energy sources, namely solid, liquid, and gaseous forms of fossil fuel constitute the largest proportion of the global energy mix. Historically, energy mix transitions have not been abrupt affairs although the pace of the transitions accelerated in the 20th century. Invariably the transitions have been associated with technological developments such as the harnessing of steam power with coal usage and the advent of the internal combustion engine with oil. In terms of energy conversion, the practical realization of electro-magnetic machines and the subsequent development of electrical power in the late 19th century [2] have provided the *end-use* energy form that is most convenient and flexible and so far available. The same electrical appliances, i.e., refrigerators, washing machines, and cookers can be used anywhere in the world if a source of electricity is available, although some slight modifications may be required to account for local supply characteristics such as voltage, current, and frequency.

What could bring about significant transitions in the present energy mix and why? In the past, cost and availability have been the prime, if not the only, drivers of energy mix transitions. While these two factors will remain *drivers* of the choices made in energy source exploitation and power production, other factors such as global population increase, urbanization, the growing human demands for clean air and clean water, issues associated with climate variability, and problems attributed to anthropogenic climate change will also play key roles in determining what present and future global societies will embrace as being a reasonable mix of energy source utilization. The emotional and physiological aspects of energy use choices should also not be ignored, but treated with caution.

For example, the "fear" of using nuclear reactor energy, is regularly claimed to have prevented [3] its more widespread adoption, but not all agree, citing other factors such as lack of technology access, initial infrastructure costs, and concerns associated with thermonuclear weapons, as the primary reasons for the lack of implementation [4]. Another example is that some elements of the global society prefer cooking with wood [5] as opposed to gas or electricity because of the imparted flavor and the maintenance of traditions. How we use the supplied energy also effects the proportions of the specific energy sources in the energy mix, locally, regionally, and globally. In Canada, more electricity is generated using hydropower [6], while in the United States the burning of coal and natural gas [7] are the main energy sources for electrical power production. These differences in energy source choice to some extent reflect the natural resources of the geographical region or country.

1.1　*Energy Sectors*

Despite the differences in energy source use the global end-use of energy and power is similar in purpose if not scale. These "purposes" are usually described as sectors, e.g., transportation and industry. The precise definition of what constitutes a sector tend to vary between countries and international agencies and organizations. In the main, throughout this chapter, the definitions used by authoritative sources such as the International Energy Authority[a](IEA), the United States' Department of Energy (USDOE) [8], the United Nations DES Statistics Division

---

[a] An international body made up of 30 member countries and 8 associated countries; https://www.iea.org/

(UNSD) [9], and non-governmental independent sources, e.g., the international energy conglomerate, BP p. l. c. (BP)[b] will be used.

Generally, while the same descriptions of the main energy-use sectors, i.e., transport, industry, and buildings are commonly used by the major data reporting agencies, there are slight differences in the precise definition of the constituent parts of a sector. For instance, the USDOE's "Energy Information Administration" (EIA) separates residential and commercial building sectors with the former only including private household and precluding "institutional living quarters," whereas BP includes all buildings. The BP sector definitions [10] are summarized in Table 1 as an illustration of sector specifications.

Table 1: BP Energy Sector Definitions [10]

BP Energy Sector	Sector Definition
Transport	Energy used in road, marine, rail and aviation.
Industry	Energy combusted in manufacturing and construction and in the Energy Industry including pipeline transportation and for transformation processes outside of power generation.
Buildings	Energy used in residential and commercial buildings plus agriculture, fishing, and in IEA's nonspecified *other* sector [8].
Noncombusted	Fuel that is used as a feedstock to create materials such as petrochemicals, lubricant, and bitumen.
Power	Inputs into power generation (including combined heat and power plants—CHP)

Each sector can use a variety of energy sources but usually some will be preferred over others. The choice of preferred

---

[b] https://www.bp.com/

energy source may change over time. In the transportation sector, coal would have been the most used energy source for the steam-driven marine and rail transport in the first half of the 20th century, but then was quite rapidly replaced by diesel fuels. Although diesel-fuelled propulsion still dominates marine transportation [11], the electrification of railways, by no means a recent innovation, has gradually replaced main line diesel-powered locomotives [12] now accounting for some 25%, by length, of global rail networks. The electricity can be generated by different energy sources.

1.2   *Energy Sources*

At the dawn of civilization, the only energy source exploited was muscle power provided by humans and then augmented by similar sourced power from domesticated animals. As humans learned how to harness the properties of fire for cooking and heating, and then to be purposely able to start fires, wood, and other combustible vegetation became increasingly the major sources of energy. Indeed, these "traditional biomass fuels"'dominated the energy mix until the opening years of the 20th century. That is not to say that these biofuels were the only source of energy as humans discovered how to use wind and waterpower and to construct more efficient methods of exploiting muscle power. Similarly, the history of solar power utilization did not start with solar photovoltaic (PV) panels, as often imagined, but at least two millennia ago when the Greeks and Romans used reflective materials—so called burning mirrors [13]—to light fires and for rudimentary magnifying lenses for the space heating of buildings. Geothermal energy in the form of natural pools and hot springs is known to have been used for cooking, space heating, and bathing for at least 10 millennia [14]. All these energy sources are still available today and likely will be for the rest of this century, and probably beyond, as they are considered "renewable." However, while an energy source may be classified as renewable, the way in which it is used may be defined as environmentally unfriendly, such as large-scale hydroelectric production[c] mainly because of greenhouse gas (GHG) emissions and habitat degradation.

When considering what constitutes a reasonable mix of energy sources there are then several candidates, old and new and yet to become technically mature, as summarized in Table 2.

Table 2: Energy Sources that Could Be Used in Global Energy Mixes

Used Prior to the 18th Century	Added Since the 19th Century	Yet to Become Technically Mature
Traditional biomass	Solid, liquid and gaseous fossil fuels	Hydrogen
Solar	Nuclear fission	Nuclear fusion
Water	Nontraditional biofuels	Ocean power
Wind		
Geothermal		
Muscle power		

The shares of major energy sources use (1980–2000) were estimated by Smil [15] and are shown in Fig. 1. The current shares are shown later as Fig. 3.

1.3 Energy Convertors

Until the arrival of utility-scale electricity toward the end of the 19th century, energy sources were used mainly for heating, cooking, and to a lesser extent, lighting and motive power. The latter gave rise to the term *prime mover* that usually referred to the generation of mechanical power from a certain energy

---

[c] The definition of large scale varies from country to country and in some instances, e.g., China, all hydropower is regarded as renewable.

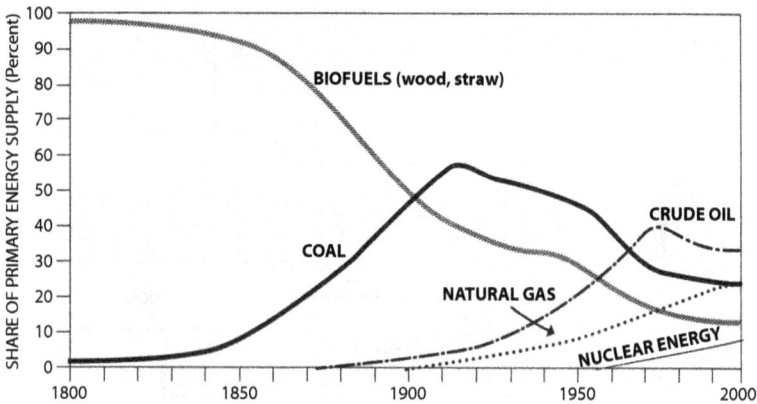

Fig. 1: Major Energy Source Shares from 1800 to 2000
[10,15]

conversion device such as a reciprocating steam engine (piston-cylinder) or a waterwheel (rotary). The generation of electrical power normally requires rotary motion and the advent of the turbine in the 19th century invented by British, American, and Swedish engineers [16] introduced new forms of prime movers now known collectively as *turbomachines*. Initially, the turbines were powered by generated steam or water flows, but since the mid-20th century hot gases have also been used to power a turbine—the gas turbine.

Once obtained, the mechanical energy can be converted into electricity using a generator or used directly in land, sea, or air propulsion systems. If the mechanical prime mover receives its energy from a thermal source, then its efficiency will be limited by Carnot thermodynamics, except when the thermal energy is transferred to the prime mover solely by radiation [17]. The thermal source could be chemical, nuclear, geothermal, solar, or oceanic temperature differences. However, wind and water energy (gravitational) can also be used to power mechanical devices in which case the efficiency will be limited by the so-called Betz Law [18]. To produce electricity, nonmechanical pathways are also available, such as fuel cells energized by chemical energy and solar-powered PV and thermoelectric devices. In general, prime movers are then rotary or reciprocating devices, but can be a combination of both, such as in radial aircraft engines. An

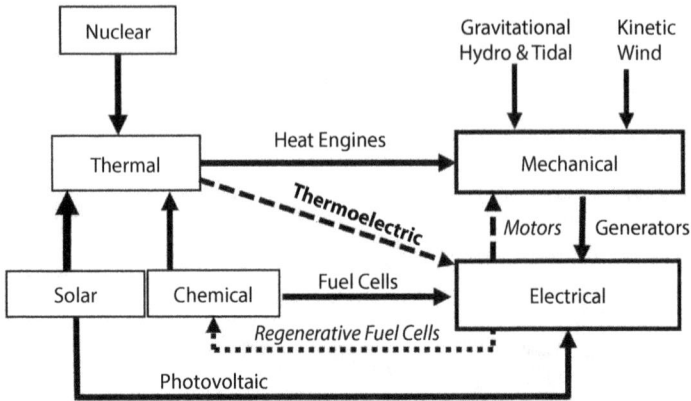

Fig. 2: Energy Sources and Convertors

inclusive, but not exhaustive, overview of energy sources and convertors is given in Fig. 2.

1.4 *Energy Mixes*

With a variety of energy sources and energy convertors available, governments, until the later years of the 20th century, would be largely responsible for regulating their nation's energy mix. This isolation of energy use decision-making began to change with the advent of an international environmental treaty to eliminate ozone-depleting substances and protect the ultraviolet radiation absorbing properties of the stratospheric ozone layer [19]. This treaty was known as the Montreal Protocol [20] and proved to be the forerunner of another international environmental treaty, the United Nations Framework Convention on Climate Change (UNFCC) [21] aimed at preventing, *dangerous anthropogenic interference with the climate system"*[22]. In 1992, UN member countries adopted the framework and, subsequently, the global responses to climate change were embodied in the "Kyoto Protocol" [23] of 1997, which came into force in 2005 and, by the start of the current decade, has 192 signatories. More recently, the 2015 "Paris Agreement" [24], which is essentially an action plan to combat the threat of human-driven climate change and in doing so to keep the *global temperature rise this century well below 2 degrees Celsius above pre-industrial levels and to pursue efforts to limit the temperature increase even further to 1.5 degrees Celsius* [21]. By February 2020, the agreement had been ratified by 187 parties.[d] A key part of the action plan of the agreement is *to accelerate and intensify the actions and investments needed for a sustainable low carbon future* [21].

However, national commitments to, and ratifications of, such treaties are voluntary and cannot be enforced by agencies external to the countries concerned. Indeed, it could be said that the only leverage institutions such as the United Nations (UN) have now is persuasion, including "naming and shaming" those countries that do not fulfil their commitments. Although "global governance" by the UN, as enunciated in a UN policy document in 2014 [25], could possibly address some of the enforcement issues, it is highly unlikely that all, or even most, countries will wholly relinquish their sovereignty to a supranational body. Nevertheless, have these unenforceable voluntary commitments impacted the global energy mix? The answer appears to be "perhaps," as illustrated in Fig. 3 showing the changes in global shares in primary energy consumption [10] by source for the period 1992–2018. A more definitive conclusion would require knowing how fluctuating oil prices and technical innovations have impacted the mix. Yet it can be observed that the shares natural gas and renewable energy sources have increased, while those of fuel oil energy and nuclear energy have decreased even though total nuclear energy production has been increased since 2012.

It is clear from the available global energy consumption data, which the challenges of achieving the objectives of the Paris Agreement, especially the realization of a sustainable low carbon future, will vary with geographic region because of differing energy mixes, such that it is unlikely there will be a global one-size-fits-all solution. For example, in 2018, the Asia–Pacific region's dominant energy source was coal, while in the Middles East and the Commonwealth of Independent States (CIS)[e] natural gas held the

---

[d] Country, State or other recognised entity such as The European Union
[e] Russia plus eight other post-Soviet Union States: Armenia, Azerbaijan, Belarus, Kazakhstan, Kyrgyzstan, Moldova, Tajikistan, and Uzbekistan. http://worldpopulationreview.com/countries/cis-countries/

largest energy source share. In all global regions, fossil fuel energy sources in their various forms, solid, liquid, and gas, still account for the bulk of the energy mix, and only in the Americas, and especially Europe, have renewable energies made measurable in-roads into the prime energy consumption markets [10], as shown in Fig. 4.

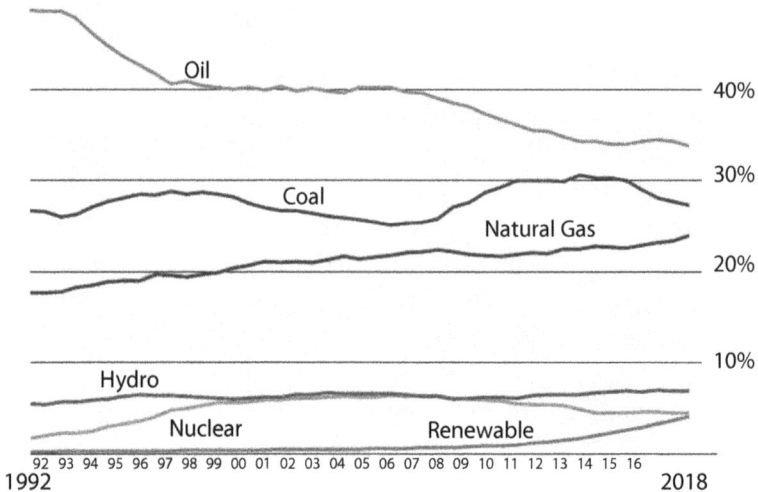

Fig. 3: Historic Shares in Primary Energy Consumption by Source [10]

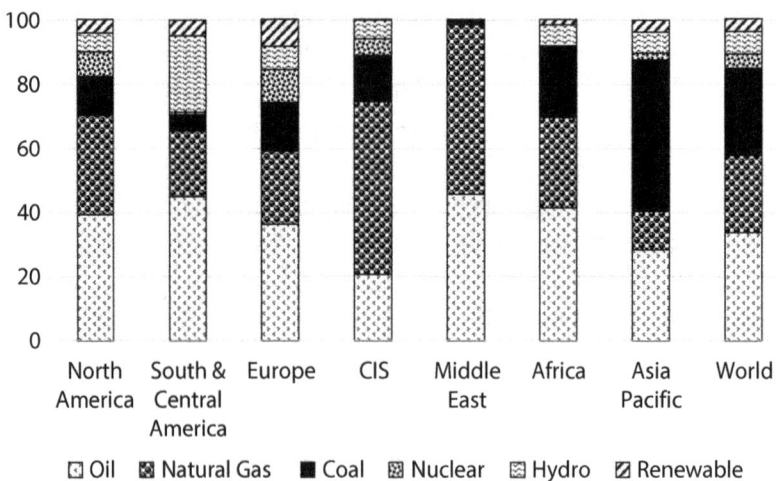

Fig. 4: Primary Energy Consumption by Region. Produced using data from Ref. 10.

As mentioned in Section 1.1, energy sources are used in specified sectors and although, as emphasized, the precise definition of a sector may vary slightly, the agencies gathering energy statistics provide meticulous details of their definitions. Table 1 is an example of this approach. Global data on energy uses by sectors are collected, analyzed, and published by the agencies citied in Section 1.1 of which, arguably, the most certified is that presented by the UNSD [26,27]. Using the data presented in Table 76 of 2019 UNSD's free Energy Statistics pocketbook, it can be seen, in Fig. 5, that the global industry sector uses more energy than the other main sectors, although transportation is a close second.

However, once again there are significant regional variations hidden in the global data as manifest in Fig. 6 showing the sector energy-use proportions of the top five final energy consuming countries [28]. It is noticeable that, among these top five users, the transportation sector energy use in the United States is significantly higher than in the other four countries and greater than the United States' industrial sector. As the dominant energy sources in the transportation sector are refined crude oils such as diesel and gasoline, by a sizeable amount, then if they are to be replaced the task would be immense, particularly in the United States.

Sector Energy Consumption - Exajoules

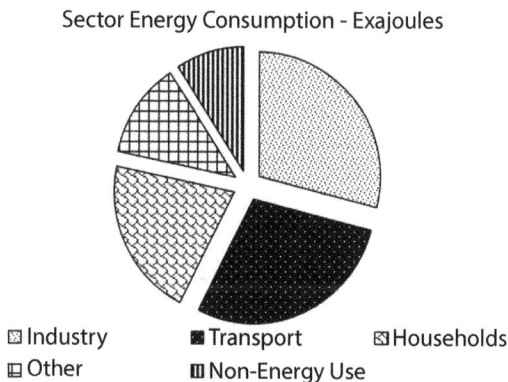

☒ Industry　■ Transport　☒ Households
▣ Other　⬛ Non-Energy Use

Fig. 5: Energy Consumption by Energy-Use Sector Using Data From Ref. [27]

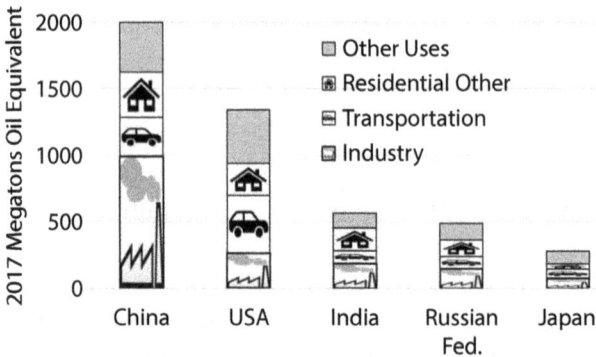

Fig. 6: Energy Consumption by Sector in Top 5 Energy Using Countries; Data Taken From Ref. [28]

1.5 *What is a Reasonable Global Mix?*

With so many different energy sources, and energy convertors available, and regional variations in how they are used in the major energy sectors, are there any general principles which could be applied in seeking a reasonable global mix of energies? Perhaps a dictionary meaning of what is "reasonable" could provide a suitable starting point? All the leading international dictionaries define the word in an essentially similar manner, so quoting *The Cambridge English Dictionary* [29] the three elements that constitute the meaning of "reasonable" are as follows:

a. Based on or using good judgment and, therefore, fair and practical
b. Acceptable
c. Not too expensive

All these facets, to a large extent, appear in the UN's Sustainable Development Goals (SDGs) [30], adopted by the General Assembly of the UN following the Paris Agreement, which collectively are a *blueprint to achieve a better and more sustainable future for all*. In total, there are 17 SDGs with SDG 1 being the most laudable of all the goals, *End poverty in all its forms everywhere*. The most relevant one for a reasonable global energy mix being Goal 7, target 7.1, namely to *Ensure access to affordable, reliable, sustainable and modern energy for all* [30]. Is the achievement of SDG 7 then the recipe for a reasonable global or regional energy mix?

2.0   Affordable Energy for All

The global population has rapidly increased since the dawn of the industrial revolution over 200 years ago, rising from 1 billion to 7.7 billion [31] and, despite a forecast decrease in growth in the UN's 2019 population report [32], it is expected to rise to 9.7 billion by 2050, and reach 10.9 billion by 2100. Much of this growth will occur in the poorest countries. Will they be able to afford *"sustainable and modern"* energy? Can they afford their current energy mixes? According to estimates from a World Bank (WB) report in 2018, 3.4 billion people, 44% of the global population, *still struggle[s] to meet basic needs* [33]; estimates echoed in a recent article in the Stanford Social Innovation Review about *the true extent of global energy poverty* [34]. In Canada, a rich, high-income country in global terms, between 6% and 13% of regional populations also experience *fuel poverty*, defined as households having to spend more than 10% of their income on energy [35].

The terms *energy insecurity* [36, 37] or *energy burden* [38] are used also to describe energy poverty by some organizations and, in the United States, by congressional politicians. Following the USEIA's Residential Energy Consumption Survey (RECS) in 2015 an analysis indicated that a third of US households face a challenge in meeting energy needs [39], as illustrated in Fig. 7. The "energy burden" is defined as the percentage of a household's income spent on energy. If this burden to income proportion is greater than 6%, then the burden is deemed to be unaffordable [39]. According to data gathered from the latest US Census, the average energy burden is 2.7% with variations in individual states from 2% to 3% excepting Maine and the territory of Puerto Rica at 4% [40]. So, there appears to be no problems with energy affordability in the United States, but these average figures hide

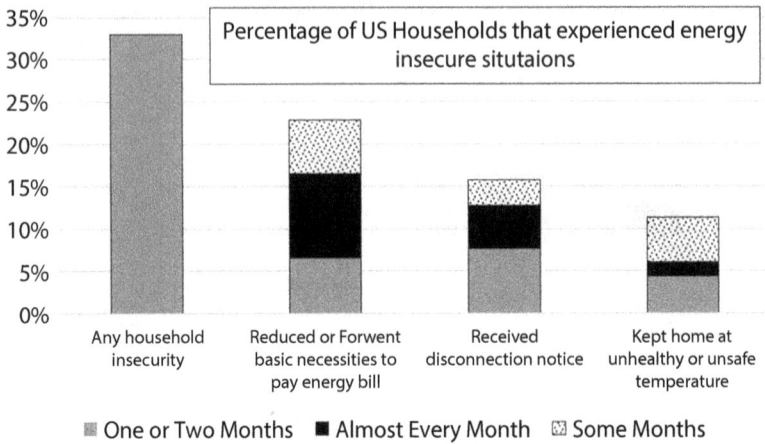

Fig. 7: US Households Facing Energy Challenges in 2015: Data extracted from Ref. [39]

the challenges experienced by low-income households, which constitute 44% or 50 million of the national housing stock [41], and on average have an energy burden of 8.6% or three times the national average. For those living on the defined Federal Poverty Line (FPL), which is US$26,200 for a family of 4, the average energy burden is 17% [42]. In the Murkowski and Scott study [37], it was estimated that if household energy costs rose by 10%—which could be said to be *not too expensive*—a further 840,000 would be pushed in poverty. A trend hardly encouraging for the noncontroversial SDG goal 1 to end poverty.

Measured by Gross Domestic Product (DGP), the United States is still the richest country in the world in 2020 [43], although, in terms of GDP per capita, several countries have higher values, but their combined population is only about 7% of the United States'. If countries like Canada, and especially the United States, cannot provide affordable energy *for all*, is it reasonable and realistic to envision that all other nations can achieve such a lofty goal? The scale of this problem is exacerbated by actual global poverty. In the United States, the FPL is US$35 per day, while the comparable International Poverty Line (IPL) is US$1.90. Based on the latter figure, the UN estimates that 42.4% of the population of Sub-Saharan Africa (410 million people) live below the poverty line, as do 30.2% of landlocked developing countries and 35.2% of the least developed countries. While the IPL is probably artificially low and needs revision [44], maybe to US$7.40 per day [45], such a higher level would still be far less than the Canadian and United States poverty thresholds.

2.1 *Access to Electricity*

Despite measures such as the FPL and IPL for actual poverty, there are no universally accepted definitions of energy poverty, burden, or affordability, which makes it difficult to accurately assess global energy poverty rates or even determine a baseline. However, a proxy measure used by such organizations as the Intergovernmental Panel on Climate Change (IPCC), the IEA, the WB[f] Group, and the World Health Organization[g] (WHO), is the lack of access to electricity, citied as the main cause for energy poverty. In 2017, according to WB [46], almost 10% of the global population did not have access to electricity, which translates to 840 million people [47]. This represents a reduction of 445 million since 1997, a commendable achievement no doubt, but how long will it take to reach 100% accessibility? At the present rates of population increase and global electrification, it could be the early 22nd century, about 2125, before complete accessibility for all is attained.[h] This estimate may appear pessimistic, but the IEA in their 2019 report on SDG 7 [48] projected that by 2030 there will about 620 million without access to electricity, which will increase to 740 million by 2050 (~8% of the global population) based on *existing policy frameworks* and *only specific policy initiatives that have already been announced*.

Where are the current 840 million? Of the countries and economies reviewed by the WB in 2017 [46], almost ⅔ have at least 97% access to electricity and of the world leading economies

---

[f] https://www.worldbank.org/en/who-we-are
[g] https://www.who.int/about/who-we-are/frequently-asked-questions
[h] Reader, G.T. (2020) Calculation based on UN and World Bank data on present rates of global population and electrification growth.

represented by the G20[i] only South Africa is below the global access average at 84.4%. However, percentages can sometimes be deceptive. Although the urban population of India has a 99.2% access to electricity, its rural population only has 89.3% access and as 66% of the population lives in such areas, then between 95 and 99 million of India's inhabitants do not have access to electricity. The neighboring states of Pakistan and Bangladesh have at least 62 million and 19 million, respectively, without access. Together these three countries account for just over 20% of the total global population without electricity access, so where are the rest? The answer is largely in Sub-Saharan Africa [49] where geographically 46 of the 54 African countries are located. According to the same WB report, already citied, 44.6% of people in Sub-Saharan Africa do not have accesses to electricity, some 463 million people. Population growth rate in this region is over twice the global average and a further billion people could be living there by 2050 according to the UN [50]. Obviously, if 100% global access is ever to be achieved, then the electrification of Sub-Saharan will be of paramount importance.

2.2   *Why Electricity?*

Electricity could be rightfully described as the *Lingua Franca* of energy. It can be generated and used in so many ways, which is its main attraction. Moreover, several electrical generating devices can be used in combination to power a single electrical motor or provide an electrical supply—a *multihybrid* system where each device can be operated at it most advantageous in terms of efficiency and power output to meet the required energy demand. In the maritime community such an approach is evolving from Integrated Electrical Propulsion (IEP) systems to Integrated Electrical and Electronic Power Systems (IEEPS) [51], which are mini grids. However, the concept has long been in use in regional and national electrical utility grid and transmission systems. The

---

[i] What is the history of the G20? https://g20.org/en/about/Pages/whatis.aspx, last accessed 11 March 2020.

epithet "utility-scale" is regularly encountered when discussing or describing these systems, but what does this mean?

In 2012, the US's National Renewable Engine Laboratory (NREL) [52] defined utility-scale for solar-powered electricity installations as being 5 MW, while the US government's Office of Energy Efficiency and Renewable Energy (EERE) typically define renewable energy projects as being 10 MW or larger [53]. These definitions have as much to do with eligibility for incentives and subsidies as with functional use. Consequently, other definitions start at only 1 MW [54]. However, if utility-scale is determined by, say, the number of households that can be powered by 1 MW plant then if the plant is operating 24 hours per day this would be enough to provide power to over 1,200 households in the Canadian Province of Alberta [55], although the same plant powered by solar energy with sufficient energy storage would only power about 200 homes.

Annually, the average household of US State of Louisiana [56] uses twice as much electricity as Albertan households, and Canadian households use 650% more than the global average [57]. In defining what is utility-scale then it appears that "size does not matter" when operating function is considered. Thus, it is perhaps more helpful to describe access to electricity as *Grid* and *Off-Grid*. Even within this seemingly simple classification there will be variations as grids can be international, national, regional, or local, as well as *distributed* and/or *smart* [58]. There can also be local *mini or micro grids*, as with the IEEPS system, where a small number of households can combine and share their electrical generation facilities such as solar PV panels or a larger village can share electricity produced by a single in-situ generating device. Initiatives, such as the ABB's (ASEA Brown Boveri)[j] "Access to Electricity" program [59], which started 2 decades ago have led to the installation of micro and mini grid systems in remote areas of India and Sub-Saharan Africa using solar panels and diesel generators. ABB have worked in conjunction with local and national governments and NGOs

---

[j] A Swiss-Swedish Multinational Fortune 500 company; https://new.abb.com/about

(nongovernmental organizations) such as the World Wildlife Fund (WWF).

These and similar initiatives have brought lighting and power for small appliances, such as cellphones and sewing machines, to some rural communities, especially since the advent of *off-grid* "pay-as-you-go" solar panels [60], where households pay for the cost of a solar panel over a fixed period of time after which they own the panel [61]. The concept has also been introduced for small business in large African cities such as Nigerian Lagos [62]. Such creativity is to be admired, and the private–public partnerships such as ABB's many projects are commendable, but they have only made very modest contributions to the global electrification problems, especially in rural areas.

A cautious expansion of the access to electricity programs in Sub-Saharan countries may be a reasonable approach if only to avoid the "Ghana-problem" of unintentionally generating excess electricity in its rush to achieve 100% access in a very short timeframe [63]. What is the Ghana-problem? Ghana's Government, to avoid another economic downturn caused, in part, by a shortage of electrical power generation in 2014, embarked on a rapid expansion of its electrical grids with private power producers and providers on a contractual "take-or-pay" basis. This means that even if the installed electrical capacity is not wholly consumed it still has to be paid for and in 2018 the unused 2,300 MW cost the Government over US$500 million. The Government also contracted liquid natural gas supplies on a similar basis and the annual excess gas capacity charges could exceed the cost of the excess electricity [62].

So, while Ghana has made tremendous progress in providing access to electricity for its population in a very short time [64], with 84% having access in 2018 and a forecast 100% by 2020 [65], it has come at an almost unbearable and unsustainable financial cost. Can contracts be broken or renegotiated? In Ghana's case, there will need to be some actions taken, but, in general, their problems emphasize the need for robust and transparent contractual processes when governments seek to improve access to electricity with independent power producers. Nevertheless, such setbacks are probably to be expected when energy transitions are attempted, and it is unlikely that they will provide unsurmountable obstacles to the global focus on increasing electrification.

3.0   Reliable, Sustainable, and Modern

3.1   *Sustainability*

Although the terms reliable, sustainable, and modern appear frequently in the energy and environment literature, as a review of the definitions of modern energy found, *there is not always complete agreement or clarity within the literature about how modern energy services are defined* [66], and this conclusion could also be applied to the definitions of sustainable [67] and reliable [68]. The sustainable definition frequently used in UN publications comes from a 1987 UN report commonly referred to as the "Brundtland report" or "the Brundtland Commission report [69]" named after the Chair of the commission Gro Harlem Brundtland, Norway's first female Prime Minister. Sustainable development is defined rather than sustainable energy, but the definition is equally applicable to all forms of sustainability, i.e., *meets the needs of the present without compromising the ability of future generations to meet their own needs* [70]. By "future generations" is it meant to the end of time, i.e., when the Earth ceases to become habitable as the sun moves toward our planet? Using a recommendation for what constitutes a generation [71] and an estimate of how long the human species can remain on Earth [72] that could be tens of millions of generations from now. Perhaps it would be more reasonable to paraphrase a version of the definition provided for renewable energy by Student Energy [73], [sustainable] *energy is produced from sources that do not deplete and can be replenished in a human's lifetime.*

According to WHO data [74], the average global life expectancy in 2016 was 72 years, but there are significant inequalities between rich and poor countries [75]. In 2019, the population of the Central African Republic has the lowest life expectancy at 53 years, while for the people of Japan it was closer to 85 years, but the UN's projection is for global life expectancy to be over 80 years by 2099 [76]. Therefore, when considering the sustainability aspects for reasonable energy mixes and energy transitions, the period 2020–2100 would be more in harmony with the Student Energy definition, but the need to use energy sources that "do not deplete" could be challenging, as a pedantic interpretation would lead to the conclusion that no such sources exist. Yet, over the period chosen, the depletion of solar energy is a nonissue and similarly with wind energy. Other renewable energy sources such as geothermal, managed biofuel–biomass energy, gravitational energy—manifest in ocean tidal power and some terrestrial hydropower—will also not be measurably depleted in the remainder of this century. All of these could play a sustainable role in energy mixes, but what about nonrenewable sources such as fossil fuels and nuclear energy?

There is no doubt that the proven reserves (current economic extraction) of fossil fuels are being gradually depleted [77] but the known resources are estimated to be orders of magnitude greater than the reserves [78] although there are concerns about how proven reserves and resources are determined [79]. Arguably then fossil fuels could meet the requirements of both the Brundtland and Student Energy sustainability definitions at least until the end of the century but during this period, it is unlikely sustainability will not be a decisive factor in including fossil fuels in energy mixes.

Although nuclear energy is usually described as nonrenewable the portrayal is contentious as although the material used in nuclear plants is not renewable, nuclear energy itself can be considered as renewable [80]. Moreover, the US Department of Energy (DOE) also considers nuclear energy to be sustainable [81] and is leading the Nuclear Innovation: Clean Energy Future (NICE Future) project in conjunction with Canada and Japan with several other countries participating such as the United Kingdom and Russia whose aim is *to ensure that nuclear energy receives appropriate representation in high-level discussions about clean energy* [82]. As with fossil fuels, it would be remiss not to include nuclear energy in energy mix and transitions discussions, although other factors may preclude or at least diminish its use.

3.2 *Reliability*

Comments abound in the literature that if the sun "does not shine" and the wind "does not blow" then no solar power and wind power is produced and so it is implied these energy sources are unreliable. While such utterances are sweeping oversimplifications as if enough energy storage capacity were to be available, the unreliability claims would be largely negated. But what is reliability? Solar and wind energy are going to be available for billions of years—isn't that reliable? Fossil fuels will eventually be no longer available so are these unreliable? It all depends upon what is meant by reliable and it appears that reliable and renewable are frequently used interchangeably. While the description reliable is used in much of the literature it is rarely defined in any definitive manner. Obviously if the supply of an energy source cannot be guaranteed, because it is contingent on external factors that cannot be controlled by the user, then it could be said to be an unreliable source. For those who enjoy access to electricity, then if the lights go out although the bills have been paid, they may be tempted to call electricity unreliable, justifiably so if it is a frequent occurrence. The reliability of the electricity supply may well be a good proxy for reliability assessments in the target specifications of SDG 7, especially since advocacy, as embodied by lawful acts in several US states [83], for a future where all energy will be renewable and 100% of electrical generation will come from 100% carbon-free energy sources. Nevertheless, electrical supply reliability will only serve as a universal proxy when the bulk of global energy demand is met by electricity.

In what ways could the reliability of meeting electricity demands be measured? One obvious method is to determine for how long the supply is interrupted in a defined time period. That is not quite as easy as it sounds, as a total loss of supply—a "blackout"—may only effect certain areas of a grid and in some instances suppliers, to combat unexpectedly large increases in demand, intentionally implement a "brownout" whereby the supply voltage is reduced, albeit by an almost unnoticeable amount, but this may cause equipment to fail or wholly shutdown. During a major blackout in 1965 known as the "Great Northeast

Blackout" on the east coast of the United States and in the Canadian province of Ontario [84] over 30 million people "lost" electrical power for 13 hours. This event prompted the eventual establishment of the North American Electric Reliability Corporation (NERC) [85] now an international regulatory authority covering the continental United States, Canada, and parts of Mexico, with a mission to develop and enforce Reliability Standards [86]. These regulations and standards for generating and supplying have evolved over the past 50 years and now include renewable electricity generation. With over five decades of experience involving collaboration with governments, state, provincial, and federal, owners, operators, and users, the basic tenets of how the NERC define "reliability" are worthy of consideration in seeking a measure of energy supply reliability.

What are these principles? Traditionally they have been (a) adequacy and (b) operating reliability. The definitions of the two principles being [68]:

> "***Adequacy*** - *The ability of the electric system to supply the aggregate electric power and energy requirements of the electricity consumers at all times, taking into account scheduled and reasonably expected unscheduled outages of system components.*"

> "***Operating reliability*** - *The ability of the electric system to withstand sudden disturbances such as electric short circuits or unanticipated loss of system components*".

NERC recommends performance objectives, how these should be assessed, and carries out periodic assessments, *subject to the oversight of the US Federal Energy Regulatory Commission and governmental authorities in Canada* [84]. This is a framework [87] that could well be adopted by the UN in tracking progress [88] in achieving the SDG 7.1 target by 2030. For the moment, it appears that data accrued by the WB, WHO, IEA, and the International Renewable Energy Agency[k] (IRENA) will used to determine the progress of SDG 7 targets in conjunction with the UNSD [89],

---

[k] https://www.irena.org/aboutirena

albeit without any definitive statement yet to be made regarding the precise meaning of "reliable."

3.3 *Modern*

As previously mentioned [66], there is no accepted definition of the meaning of the term *modern energy*. However, the phrase is a truncation of *modern energy services* that has been used in some recent UN publications dealing with SDGs, e.g., SDG 7.1 is described as *By 2030, ensure universal access to affordable, reliable and modern energy services* [90]. In an earlier UN report dealing with Millennium Development Goals (MDGs), the forerunners of the SDGs, in the discussions regarding energy the term *Modern Fuels and Electricity* [91] is used. But what are modern fuels and modern energy services? The energy sources available in 2020 have long been so, even the PV effect has been known for over 180 years, and the first solar PV cells were constructed over 140 years ago [92], almost a century before the so-called Third Industrial Revolution [93]. Relatively speaking, nuclear fission, discovered in the late 1930s, is a modern energy source [94], as are Ocean tidal [95] and Ocean thermal energy (OTEC) [96]. However, there are modern versions of much older energy sources such as solar, wind, geothermal, and biomass–biofuels. Generally, the modern versions use much improved energy conversion systems that are more efficient or are produced using modern techniques, e.g., renewable hydrocarbon biofuels [97].

There are no definitive statements regarding what modern energy services are, but access to electricity is unequivocally a key element of such modern services, as is access to cleaner, nonpolluting energy sources used for purposes other than electrical generation such as transportation, space heating, and cooking. The IEA [98] have identified some *significant* commonalities in the definitions of energy access, which they have linked to modern energy services and these are

a. *Household access to a minimum level of electricity*
b. *Household access to safer and more sustainable (i.e. minimum harmful effects on health and the environment as possible) cooking and heating fuels and stoves*

c. *Access to modern energy that enables productive economic activity, e.g. mechanical power for agriculture, textile and other industries.*

d. *Access to modern energy for public services, e.g. electricity for health facilities, schools and street lighting*

Although these four benchmarks are global measures, they are especially pertinent to areas such as Sub-Saharan Africa and remote communities in parts of Asia. As a key UN concern in these locations is the well-being of low-income and poverty-line households, the IEA have concentrated on the steps that need to be taken to provide access to modern energy and modern energy service as listed above in categories (a) and (b) for these households by 2040 [98]. They have specified a basic electricity need of 1,250 kWh per annum per household and access to modern fuels for cooking including improved biomass cookstoves. To get this into perspective, the average US residential customer uses almost nine times more electricity annually, 10,909 kWh [56], than the IEA baseline. Richer more developed countries and regions have already achieved goal (a), their main problems being affordability and energy poverty. To realize the other goals in terms of modern energy services, many countries have political plans to shift to renewable and sustainable energy source and reduce harmful emissions. The European Union's (EU) "super grid" concept, suggested initially in 2010 [99], is for all EU countries to have their electrical grids interconnected, together with some nearby North African and non-EU countries.

In 2019, the European Commission announced that almost US$0.9 billion would be allocated to support such energy infrastructure projects. The idea of the super grid is that when a country generates too much energy, but simultaneously another country is not producing enough, the surplus can be fed to the deficient country. This is by no means a new concept and several US states and Canadian Provinces have such arrangements and the United Kingdom has had a National Grid since the middle of the 20th century [100]. However, in the United States, states are wary of having a national grid as this could place the control of electricity generation and distribution under the Federal Government. This has led to some US States, who have embraced renewable energies, to experience "overgeneration" of solar and wind-powered electricity and, since they can do nothing with the surplus, the energy is wasted as it is not sold to a neighboring state or any regional integrated grid [101]. Canadian provinces sell 9% of their surplus generation to their US neighbors [102] and depending upon the contractual arrangements existing at the time the exported power can be provided at cheaper rates than their own consumers are charged [103].

The European super grid is aimed at overcoming these problems and by doing so encourage the increasing use of modern renewable energies as overgeneration waste will be eliminated. Thus, access to modern energy services in terms of electricity can mean access to smart, integrated, and national or regional grids. The concept can also work for mini and micro grids down to the household level [104]. The ABB projects, previously mentioned, in remote locations in Africa and India are indicative of the micro grid, household power sharing approach, but do not always provide access to modern energy services. Nevertheless, for the near future the mini or micro grid approach holds more promise for low-income developing countries than installing large national or multicountry super grids that require significant levels of initial capital investments. The scale of the problem can be appreciated by comparing the actual land size of Africa with those of the United States and the EU, Fig. 8 [105]. Mini grids for remote communities in developed countries using small- to medium -sized nuclear reactors could also become a reality in the second half of this century and augment the role already played by nuclear power in present energy mixes, see Section 3.3.1.

THE USA　　　EU + THE UK

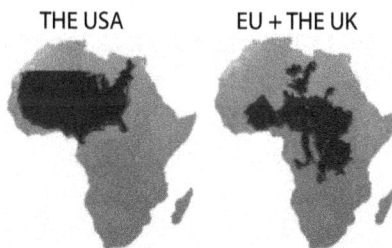

Fig. 8: Relative actual areas compared with Africa [105]

Accessing modern energy services means more than just access to electricity although that may be the ultimate and ambitious global aim by the end of this century, powered by renewables or low-carbon energy sources [106]. It appears that modern biomass–biofuels are set to play key roles in energy mixes for the remainder of this century as they are considered, somewhat contentiously, to be zero-carbon or carbon-neutral fuels and, therefore, acceptable in present climate change scenarios. Liquid biofuels can provide renewable replacements for gasoline and diesel fuels in the transportation sector. In the United States, most of the gasoline available at filling stations contains plant-based ethanol usually around 10% (E10) [107] with the bulk of E10 gasoline being sold in the agricultural Midwest states where the ethanol is produced. In countries with little or no crude oil deposits, the use of "home-grown" biofuels could become attractive for vehicular use and power generation. However, the importance of biomass in near-future energy mixes is that in 2018 almost 35% of the global population does not have access to clean-cooking fuels of which around 90%, 2.37 Billion, still use "traditional biomass" [108]. What is meant by "traditional biomass"?

Several slightly different definitions have and are used, perhaps the most detailed discussion on the topic was that by Karekezi et al. [109] but for this chapter the specification of Traditional Biomass will be that used by the IEA in the clean-cooking database [108], i.e., solid biomass (wood, animal waste, charcoal), kerosene, or coal. According to the IEA database, over 97% of the global population using traditional biomass for cooking live in Sub-Saharan Africa (848 million) and Developing Asia (1.46 Billion).[1] Once again emphasizing that if the targets of SDG 7 are to be achieved these regions of the world need to be a primary focus. Modern biomass or, perhaps more accurately, bioenergy will be a part of future energy mixes, but how much and when will be further discussed in Section 3.3.2.

---

[1] 99% of the populations living in India, Bangladesh and Pakistan, without access to clean cooking fuels use traditional biomass

There are several modern improvements in generating electricity using traditional fossil fuels in terms of energy use efficiency and lower GHG emissions, such as ultra-supercritical coal-fired power plants [110], combined cycle power plants (CCGT) [111], which use a combination of steam and gas turbines, and Allam-cycle power plants [112] where supercritical CO_2 is the working fluid rather than steam. General Electric's CCGT power plants in France (605 MW) and Japan (1.19 GW) have both achieved gross efficiencies of over 62%. A 900 MW CCGT plant planned for Canada is forecast to be completed in 2022–2023 and *is expected to produce 62% less CO_2 equivalent per MWh compared to current coal-powered electricity generation facilities* [113]. However, while these are all modern energy devices, they use carbon-based fuels and those adamant that only zero-carbon renewable and sustainable energy sources are capable of addressing climate change will not welcome the use of fossil fuel technologies in future energy mixes. So, will such modern energy devices be viewed as part of a global anthropogenic climate change mitigation strategy or as encumbrances to eliminating such climate change? For some, the former appears to be the answer, e.g., China and Pakistan are already operating sizeable *clean* coal electrical power generation plants [114].

3.3.1 Nuclear Energy

The first time a nuclear reactor generated electricity was in 1951 in Idaho, United States [115]. By 1955, Arco, Idaho, became the first community to be wholly powered by a nuclear reactor, although a larger nuclear power plant had been commissioned in the, then, Soviet Union a year earlier, and a year later the world's first "commercial nuclear power station" was opened at Windscale in the United Kingdom [116]. Today, about 440 nuclear power stations provide almost 10% of the global electricity, with a further 220 reactors (approximately) being used for research, training, and the production of medical and industrial isotopes [117]. These research reactors [118] are located globally, including 87 in developing countries, Fig. 9, and 109 of the reactors have thermal power ratings [119] of 1 MWt (thermal) or more. Although these research reactors are not used for electrical

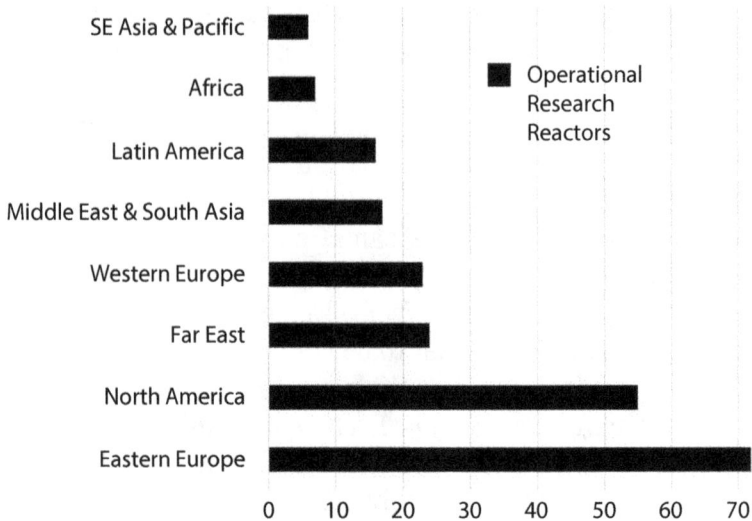

Fig. 9: Research Reactors location: Data extracted from Ref. [118]

power generation, they are indicative of the ability to construct *very* small nuclear reactors.

Other reactors, with much larger outputs than these research reactors, but still small in comparison with conventional nuclear power stations, which feed the grid, have been used for 5–6 decades to provide power for military submarines, aircraft carriers, a unique deep submergence research vehicle [120], and icebreakers. The International Atomic Energy Agency[m] (IAEA) define small- and medium-sized reactors (SMR), with *small* generating less than 300 MWe (electrical), and a *medium*-sized reactor producing up to 700 MWe [121]. However, the acronym SMR is now frequently used to describe "small modular reactors," which can be used as stand-alone units or, collectively, to form a larger nuclear power plant [122]. Several countries: Argentina, Canada, China, the United States, and the United Kingdom are taking more than just a passing interest in small, modular reactors aimed at providing energy to rural and remote

---

[m] IAEA: https://www.iaea.org/

communities; some would be *very small* with a size of about 15 MWe [116].

Natural Resources Canada (NRCan) have developed an SMR roadmap [123] that identifies three main applications: (a) On-grid power generation for coal replacement; (b) On- and off-grid combined heat and power (CHP) for heavy industry; and (c) off-grid power for remote communities, and estimate that the global market for such reactors could be $150B CAN between 2025 and 2040. However, China is taking the lead in SMR development with a 210 MWe plant using two reactors already under construction [116]. Nuclear energy then is set to be part of many countries' energy mixes and, if the technology is licensed at reasonable rates, the impact in Africa and Asia could be significant. Moreover, by the late second half of the century, the utopian promise of nuclear fusion may be finally fulfilled. In the United Kingdom, the government have announced an investment of £200 million to deliver fusion electricity by 2040 with demonstration models operational by 2024 [124]. This is a very ambitious timeline, but not as improbable as it may appear [125].

3.3.2 *Modern Bioenergy*

Biomass and bioenergy, traditional and modern, are deemed to be carbon-neutral since they are plant and vegetation based and, therefore, only releasing captured carbon dioxide, which itself will be eventually recaptured, sequestrated through new plant growth. However, the sequestration will take a finite amount of time, called *carbon payback* [126]. The payback time can vary between 1 and 1000 years [127], although the lower end of that period appears to be the more likely [128], but burning wood from Norwegian boreal forests can result in a payback delay of 150–230 years [129]. Nevertheless, as biofuels could be used to displace conventional coal, oil, and gas fuels there could be a net positive impact on carbon dioxide emission levels during the payback time periods. The situation is by no means clear, which has led to some discomfort about the carbon-neutrality of biofuels [130], but not all political policymakers have such qualms [131,132].

Crops and waste materials such as municipal solid wastes (MSW) can be used as feedstocks for conversion to liquid, gaseous, and solid biofuels [133], Fig. 10. Methane gases emanating from landfill sites can also be purified to produce so-called renewable natural gas [134]. The carbon-neutral labelling for biofuels was almost sacrosanct until relatively recently, but is now being challenged [131,135]. Regarding agricultural crops, the food versus fuel debate will continue and it could be that some bioenergy products will lose their carbon-neutral moniker, but not their place in future energy mixes.

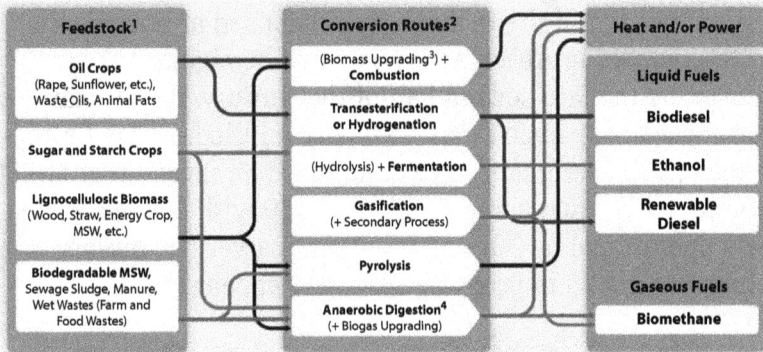

Notes: 1. parts of each feedstock, for example, crop residues, could also be used in other routes, 2. Each route also gives co-products. 3. Biomass upgrading includes any one of the densification processes (pelletization, pyrolysis, etc). 4. Anaerobic processes release methane CO2 and removal CO2 provides essentially methane, the main component of natural gas; the upgraded gas is called biomethane.

Fig. 10: Biofuel Production Pathways Ref. [133]

4.0 Anthropogenic Climate Change

In 1992, UNFCC defined [136] "Climate Change" as meaning *a change of climate which is attributed directly or indirectly to human activity that alters the composition of the global atmosphere.* These changes are *in addition to those caused by natural climate variability over comparable time periods.* The adjective "anthropogenic" is frequently not used, so an individual or organization who may believe that our planet's climate is changing, but not because of human influences, or at least not solely due to such influences, can attract the disparaging and unhelpful epithet, "Climate Denier." To determine the anthropogenic impact, a series of climate models have been developed to predict, or project, how, in the future, the two key measures of changes in climate, namely average global surface temperature, and average global sea level, rise.

As with all models they can provide an insight into the factors involved in system changes, but their predicative abilities are less robust. Ideal thermodynamic cycle model analyses, e.g., the "Otto" cycle [137], demonstrated that the thermal efficiency (fuel economy) of a specific type of reciprocating heat engine depends upon the volumetric compression ratio of the engine and the "adiabatic" index of the working gas. However, such analyses would not be used to either design such an engine or predict its performance although observations have confirmed that increasing the compression ratio and the adiabatic index of an actual engine do improve its performance. Models, however, do not always identify the important parameters that may not have been included in the analyses. Usually when there is a lack of correlation between model predictions and actual measured observations, ways of improving the model are sought. This type of approach is usually referred to as "the Scientific Method" whereby observations are used to test hypotheses.

For the moment, the differences between the climate model projections (of the two key measures previously mentioned) and observations are of such a scale that the models need to be improved. However, the media and political entities put more emphasis on climate model projections than observations and this can lead to precipitous and harmful policy decisions. Conversely, the model projections have raised potential future environmental issues that cannot be ignored. One of the main issues is the apparent detrimental effect that increasing levels of "anthropogenic" carbon dioxide is having on elevating average global temperatures. Observations [138] have confirmed that atmospheric CO_2 levels are rising, and that average globally sea levels and surface temperatures are rising, albeit not in all geographic locations. The main culprit in all these effects is hypothesized as being the increases in anthropogenic CO_2 generation emanating from the use of fossil fuels.

Consequently, despite the many uncertainties surrounding climate modelling [139], as climate is the statistical average of observed weather patterns over a defined period, usually 30 years

[140], then if human-generated CO_2 is the cause of changing weather patterns, action to reduce these levels must be taken sooner rather than later. This is because of all the GHSs; carbon dioxide emissions stay in the atmosphere for decades and in some measure for centuries [141]. The implication for energy mixes and transitions is the likelihood of continued global efforts to reduce the use of carbon dioxide producing energy sources and to replace them with sources that do not. However, the sensitivity of one key parameter by which climate change is modelled and predicted, i.e., average global temperature rise, to changes in carbon dioxide concentrations, may be far less than previously assumed.

Recently published studies from the IPCC [142], GWPF [143], and the American Meteorological Society [144], among others, have concluded that the "forcing" of carbon dioxide may result in temperature rises closer to the lower end, i.e., 1.5°C, of the modelling predictions of 4.5°C–1.5°C first proposed in the now famous Charney report [145] and that are often repeated. The reassessments have been made possible by more measured data becoming available. The emphasis, perhaps overemphasis, of carbon dioxide's role in actual climate change may well moderate as more data becomes available, but only in so far as the role of other contributing factors to actual changing climates need to be clarified, measured, and embedded in better predictive models. With better insights into what causes the climate to change, the steps that need to be taken for mitigation or adaption, or some combination of both, should become evident.

5.0   Possible Energy Mix Scenarios

Agencies that report annual energy use also predict or project how national energy and global energy will change in the future. To do this, the current trends in types of usage are considered as well as detailed "what-if" scenarios. A typical example is the USEIA's "Annual Energy Outlook (AEO) 2020 [146] with projections to 2050" that defines a reference case [147] and then uses the National Energy Modelling System [148] (NEMS) software package to project future uses. NEMS is a model that integrates "economic changes and energy supply, demand and prices" but EIA stresses the uncertainty of *energy markets as well as future developments in technologies, demographics and resources*[146]. The USEIA also produces an International Energy Outlook (IEO), the latest one being IEO 2019 with projections to 2050 [77]. In this case, a reference case is also defined along similar lines to the United States only outlook analysis, but considering regional markets as opposed to national markets and using another model, the World Energy Projection System Plus (WEPS+), which is described as "an integrated economic model" [77]. In both the AEO and IEO, the key uncertainty is oil price, and so high-price and low-price variations are considered, along with the other assumptions of the reference cases. Consequently, the economies of the less rich non-OCED countries are projected to grow more quickly than OECD[n] countries.

BP also provide annual reports that include projections out to 2040 and adopt a similar philosophy to USEIA in defining scenarios such as "evolving" transitions and "rapid" transitions [149]. The IEA and the IPCC, among others, also provide periodic reports dealing with actual energy use and future projections. As with all modelling projections, there are uncertainties in the analyses and the results. However, there are trend commonalities in the forecasts of all these agencies' projections, such as (a) the increasing role of renewable energy sources, especially solar and wind, (b) a relative energy share reduction, but still substantive use of coal, (c) increasing levels of electricity generation globally, (d) carbon dioxide emissions increasing with population increases especially from low-income developing countries, (e) regional differences of energy mixes becoming more apparent, (f) regional production and consumption differences so that some countries become net exporters of energy, while others import more energy resources, (g) demand for energy increasing globally, but energy efficiencies in developed countries offsetting energy intensity, i.e., the use of energy per capita.

---

[n] OECD – The Organization of for Economic Co-operation and Development. https://www.oecd.org/about/

The organizations and agencies that publish annual energy data and future forecasts provide numerous data tables, graphs, infographics and accompanying electronic presentations, and digital media. How can these be used in the inference of what constitutes a reasonable energy mix? Several trend commonalities of the published energy data were identified in the previous paragraph, but specific examples can also provide insights into how energy transitions could evolve. A theme underpinning the achievement of the targets in SDG7 are access to electricity and meeting future energy demands by increasing the use of renewable energies and their share of global markets. The steps taken to embrace these targets by larger and richer countries are possible indicators of changes that can be anticipated in energy mixes over the next three decades. The largest economy presently is the United States, but China, with the largest population, is set to overtake the United States and in turn will be supplanted by India in terms of population in the next decade or so, and in terms of national economic size by the end of the century. The implications of this situation for energy demand for electricity is illustrated in Fig. 11, covering the next three decades. The data to produce this graph were extracted from the USEIA's latest AEO

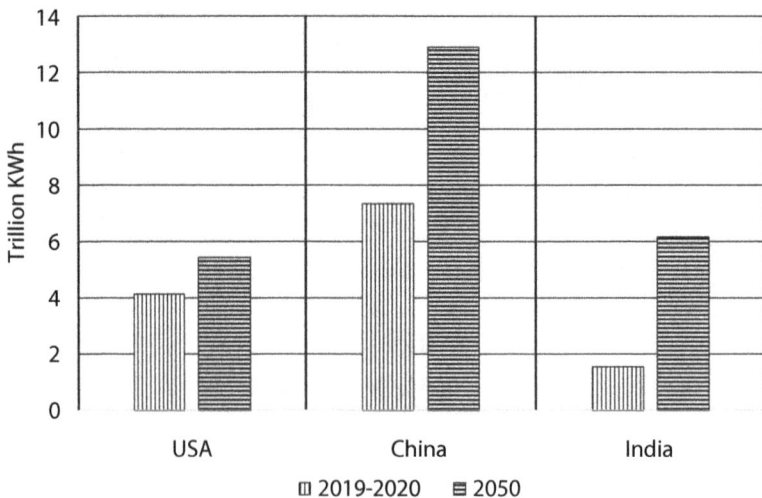

Fig. 11: Energy Demand Increases to 2050. Chart produced using data extracted from Refs. [146,149]

[146] and IEO [77] reports. Demand for energy in all countries is forecast to increase and although China's demands will almost double, India's will quadruple; by comparison, the United States demand will increase by less than a third because of lower population increases and increasing energy efficiencies.

How will these increasing demands be met? Will renewables provide 100% of the energy required not only for all electrical generation, but also for the demand increases, and eventually for all global energy consumption? The forecasts for the energy source mix to 2050 for China, India, and the United States indicate an increasing use of renewable energies for electrical generation, but measurably less than 100% and with coal (China and India) and natural gas (United States) still playing significant roles in the mixes by mid-century, Fig. 12. Nevertheless, it must be remembered that energy for electricity generation is only part of the total energy consumption mix. However, as the century progresses the amount of electricity in the mix could increase further if Electric Vehicles (EV) replace Internal Combustion Engines (ICE) in the transportation sector. When will this happen? By 2040, if other states and countries follow the lead of the Canadian Province of British Columbia, which have mandated

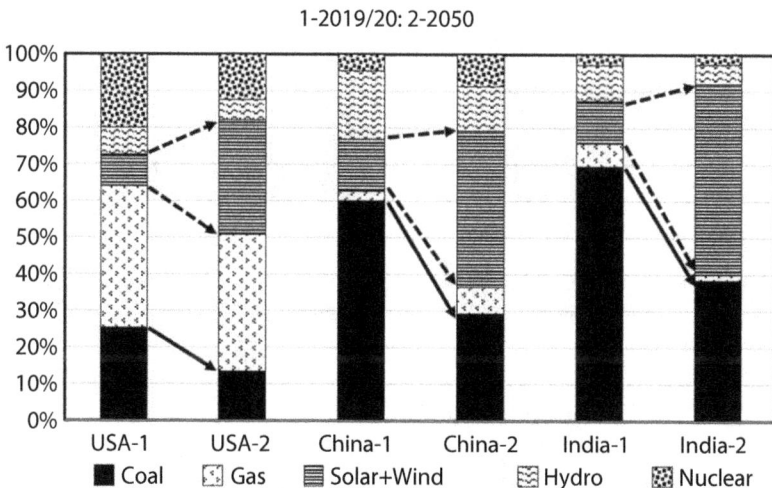

Fig. 12: Changing Energy Scenarios for the United States, China, and India to 2050: Data Extracted From Refs. [77,146]

that 100% of vehicle sales have to be zero-emission vehicles (ZEV)° by then [150]. Several other countries have similar targets, as opposed to mandates, but have indicated that sales of new ICE vehicles will be banned, in some cases within this decade.

In general, this could mean that almost all petroleum products will be replaced and if the battery charging electricity is to be zero-emission then coal and natural gas-powered electricity generation will also have to be replaced. What about the other energy sectors? At the 2019 UN Climate Action Summit [151], and emphasized in the subsequent report [152], the *The Climate Action Summit reinforced 1.5°C as the socially, economically, politically and scientifically safe limit to global warming by the end of this century, and net zero emissions* [Author's emphasis] *by 2050 as the global long-term climate objective for all.* Is this realistic? Not according to the USEIA's forecast for the global energy mix by 2050 [76] as shown on Fig. 13 and while renewables are forecast to have the largest share of energy consumption, slightly more than shown on the figure if liquid biofuels are also counted as renewable, fossil fuels still constitute a significant part of the mix.

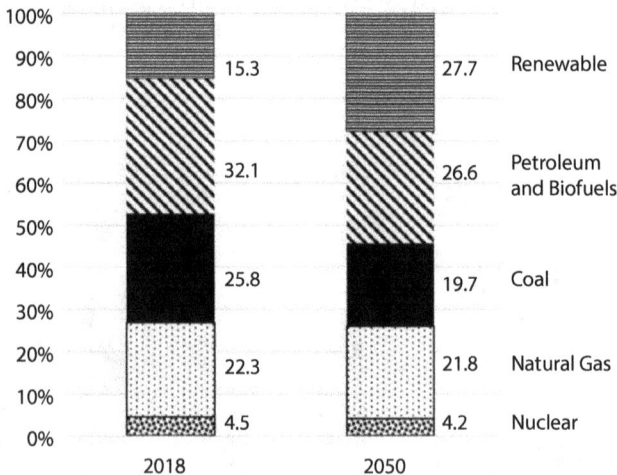

Fig. 13: Global Energy Mix Changes 2018 – 2050: Data extracted from Ref. [76]

° ZEVs include Fuel Cell Electric Vehicles (FCEV), Battery Electric Vehicles (BEVs) and Plug-n Hybrid Electric Vehicles (PHEV).

Production and consumption are not the same measure. To meet Germany's ambitious targets for renewable energy use (60% of gross consumption and 80% of power consumption) [153], the country will have to import energy from other European countries [154] presumably using the European super grid? Conversely, the United States will produce more energy products, such as liquid fossil fuels, than it will consume. Rather than leave these resources "in-the-ground" they will be exported [147] to meet the needs of other countries and regions, Fig. 14. In the concept of the European super grid, some countries will also become energy exporters, while other will become importers, such as characterized by the German approach, but it is also expected that the export–import balance for some countries will be seasonal.

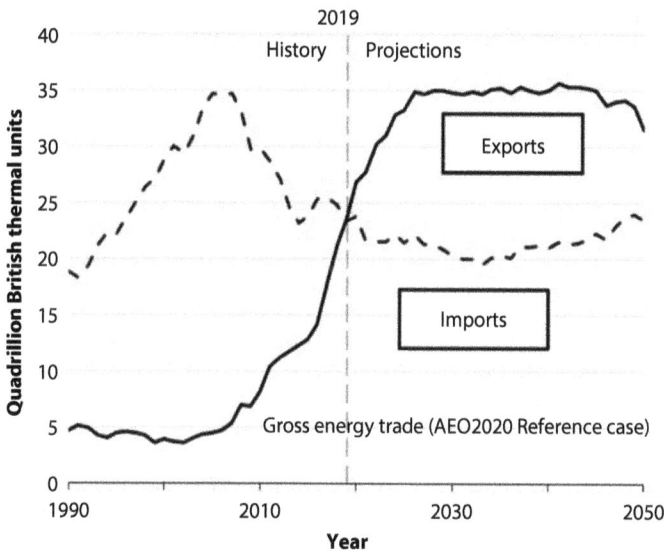

Fig. 14: USEIA projection of United States becoming energy exporter [147]

6.0　Concluding Remarks

Earlier the question was posed whether a future energy mix would be the same for regions globally or whether these mixes would be regional? From the foregoing discussion, it seems likely that regional mixes will prevail at least until 2050 and possibly to the end of this century and beyond, although these mixes will change. The consumers will play a significant role in deciding what mixes are reasonable as defined in Section 1.5, as least in democratic countries and states. Government mandates will dictate the direction of such mixes, but with ever changing Governments on 4- to 5-year cycles, mandates and targets can be altered and changed. Attitudes to nuclear energy may well change, but if so, the necessary infrastructure will not be available much before mid-century. However, with the global commitments, especially regarding anthropogenic climate change and GHG emission reduction, can it be expected that regardless of the actual energy mixes that these commitments will be met? If the 2018–2050 projections for energy-related carbon dioxide emissions, Fig. 15, are a plausible benchmark, the answer is no, as these emissions are set to increase overall, especially in lower income and developing countries.

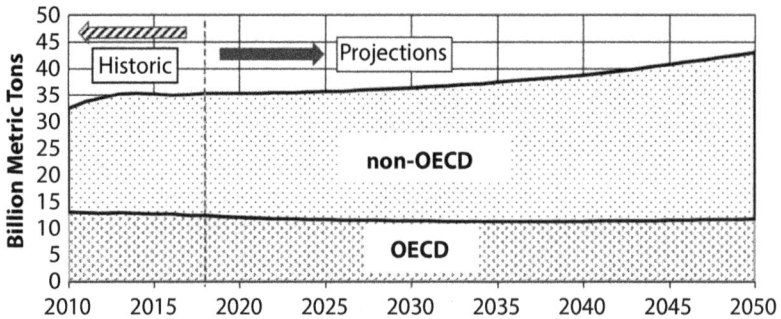

Fig. 15: Carbon Dioxide Emission Projections to 2050: Ref. [77]

The need for more energy to slake the needs of an increasing global population cannot be wholly met by fossil fuels alone, which, although dominant today, still cannot provide all the global energy demands. So, regardless of climate change concerns, other energy sources are needed now and increasingly so in the future. Increasing the uses of renewable energies have been identified as the answer to the coming energy-gap problem and as a longer-term solution as fossil fuel supplies eventually become increasingly exhausted, although it is still not clear when this will

happen. Renewables also offer the possibility of GHG emission
reduction and until the questions regarding GHG sensitivity are
answered by more observations the severity of the "greenhouse"
effect will remain the focus of environmentalists.

In the case of GHG emissions, the methodologies for the
measurements of the targets that have been set by various
international environment agreements, especially the Paris
Agreement, have led to some discussion, dissent, and obvious
anomalies in how national commitments are viewed. For example,
China, in its National Determined Contribution (NDC), has
committed to increasing its carbon dioxide emissions until 2030,
up to 35% from 2010, but also to increasing its forest stock as
part of its plan [155]. But reportedly, Canada's attempts to include
their massive forests as carbon-sinks have been rebuffed by the
UN [156]. So, while the burning of biomass and the combustion of
biofuels are defined as carbon-neutral and, therefore, not included
in the determination of carbon dioxide emission generation, carbon-
sinks such as forest regeneration and tree plantations cannot be
used to offset such emissions. This may change after 2050 as
countries committed to the Paris Agreement agreed to *achieve a
balance between anthropogenic emissions by sources and removals by
sinks of greenhouse gases in the second half of this century* [157].

Energy transitions, as previously mentioned, are usually
governed by cost and availability. It has been shown in this
chapter that not all energy sources are inherently available,
wholly or in part, to all countries and regions and therefore this
factor will have an impact on the chosen energy mix. As for costs,
metrics such as the Levelized Cost of Electricity (or Energy)—
LCOE—are often used to compare electricity generation costs
of the different energy sources and convertors, which could be
used over the lifecycle of a specific combination. With LCOEs,
the costs to build and operate a power plant are divided by the
fiscal revenues from the sales of the generated electricity [158].
As the cost of solar PV cells has rapidly decreased this century
and double-digit reductions in the costs of onshore wind power
have been achieved [159], the LCOEs (Electricity) for *new* build
solar and wind plants have been estimated to be competitive
with *existing* coal-fired and nuclear power plants and could be

lower [160], even without subsidies. All LCOE analyses obviously make certain assumptions and these involve uncertainties that occur especially when energy sectors other than electrical generation are considered. What if the cost reductions in solar and wind devices do not continue to decrease? What if conflicts and disasters lead to further significant reductions or similar increases in the price of liquid and petroleum fossil fuels?

LCOEs (Electricity or Energy) are predicated on an assumed cost of a barrel of oil or equivalent. History shows that whenever the price of oil and gas has increased, interest and investment in alternative energies have increased, but when the oil and gas prices fall the interest and investment abates. However, in this century governments have invested in renewables regardless of oil or gas prices and used incentives or legislation to ensure the development of solar and wind power, but investments require wealth, and if a country's economy (GDP) suffers a significant and sustained downturn there will not be the wealth to invest. In such cases, to continue to promote and legislate the use of renewables when cheaper traditional energy sources are available is likely to be politically indefensible. Arguments that investments, regardless of the ability to pay, will benefit future generations because of the harmful effects of the changing climate may not be sufficiently persuasive or judicious.

Nevertheless, despite all the obstacles that changing energy mixes encounter they will be overcome eventually, but any changes are unlikely to be at the pace necessary for fulfilling the UN's SDGs by 2050 or even by the end of the century. In the meantime, what could be a reasonable energy mix for some countries will not be others.

References

(1)   Smil V. *Energy and Civilization: A History*. Cambridge, MA. London, England MIT Press; 2017 .

(2)   Bowers B. *A History of Electric Light & Power*. Stevenage, UK. New York, NY Peter Peregrinus Ltd; 1982. Stevenage UK and New York

(3)   Shellenberger M. Why I changed my mind about nuclear power: transcript of Michael Shellenberger's TEDex,

Berlin. *Environmental Progress*. November 21, 2017. http://
environmentalprogress.org/big-news/2017/11/21/why-i-
changed-my-mind-about-nuclear-power-transcript-of-michael-
shellenbergers-tedx-berlin-2017.

(4)   Barnard M. Public fear of nuclear isn't why
      nuclear energy is fading. *Clean Technica*. March
      15, 2019. https://cleantechnica.com/2019/03/15/
      public-fear-of-nuclear-isnt-why-nuclear-energy-is-fading/.

(5)   Rhodes EL, Dreibelbis R, Klasen E, et al. Behavioral attitudes
      and preferences in cooking practices with traditional open-fire
      stoves in Peru, Nepal and Kenya: implications for improved
      cookstove intervention. *Int J Environ Res Public Health*. October,
      2014;11(10):10310-10326.

(6)   NRCan. *Energy Use Data Handbook*. 16th ed. 1990–2016. http://
      oee.nrcan.gc.ca/corporate/statistics/neud/dpa/menus/trends/
      handbook/tables.cfm.

(7)   USEIA. Electrical data browser: net generation, United States.
      All sectors. 2001–2019. Washington, DC: United States
      Government Energy Information Administration; 2019. https://
      www.eia.gov/electricity/data/browser/.

(8)   The United States Department of Energy. https://www.energy.
      gov/.

(9)   United Nations. Energy statistics – 2017 yearbook (United
      Nations, Department of Economics and Social Affairs, Statistics
      Division). New York, NY: United Nations, Department of
      Economics and Social Affairs, Statistics Division; United
      Nations; 2017. https://unstats.un.org/unsd/energystats/pubs/
      yearbook/.

(10) BP. *Statistical Review of World Energy*. 68th ed. Middlesex,
      TW: BP International Limited; 2019. https://www.bp.com/
      content/dam/bp/business-sites/en/global/corporate/pdfs/energy-
      economics/statistical-review/bp-stats-review-2019-full-report.pdf.

(11) United Nations Conference on Trade and Development.
      Review of maritime transport. 2017. https://unctad.org/en/
      PublicationsLibrary/rmt2017_en.pdf.

(12) SCI Verkehr GmbH. New study: railway electrification continues
      to grow in 2018. *Mass Transit*. April 19, 2018. https://www.
      masstransitmag.com/rail/press-release/12408644/sci-verkehr-
      gmbh-new-study-railway-electrification-continues-to-grow-
      global-market-development-2018.

(13) Rashed R. A pioneer in anaclastics: Ibn Sahl on burning mirrors
      and lenses. *J Hist Sci Soc*. September, 1990;81(3):464–491.

https://www.journals.uchicago.edu/doi/abs/10.1086/355456? journalCode=isis

(14) Lund JW. Geothermal energy. *Encyclopædia Britannica*. April 30, 2018. https://www.britannica.com/science/geothermal-energy. Accessed February 17, 2020.

(15) Op. Cit. Reference (1); page 376 Fig. 7.3.

(16) Meher-Homji CB. The historical evolution of turbomachinery. Proceedings of the 29th Turbomachinery Symposium; September 2000; College Station, TX: Texas A&M, Turbomachinery Laboratory, Texas A&M University. 2000:281-321. https://oaktrust.library.tamu.edu/bitstream/handle/1969.1/163364/t29pg281.pdf? sequence=1.

(17) Bejan A. *Advanced Engineering Thermodynamics*. New York, NY. Chichester. Brisbane. Toronto. Singapore John Wiley & Sons; 1988.

(18) Betz's Law. https://en.wikipedia.org/wiki/Betz%27s_law. Accessed 16 November, 2020.

(19) UNEP. *What is the Ozone Layer?* Nairobi, Kenya: UNEP Ozone Secretariat; 2020. https://ozone.unep.org/ozone-and-you#what-is-ozone

(20) United States Department of State. *International Actions – The Montreal Protocol on Substances that Deplete the Ozone Layer.* Washington, DC: United States Department of State; 2020. https:/www.state.gov/key-topics-office-of-environmental-quality-and-transboundary-issues/the-montreal-protocol-on-substances-that-deplete-the-ozone-layer/.

(21) United Nations Climate Change. What is the United Nations Framework Convention on Climate Change? 2020.https://unfccc.int/process-and-meetings/the-convention/what-is-the-united-nations-framework-convention-on-climate-change.

(22) Wikipedia. United Nations Framework on Climate Change. 2020.https://en.wikipedia.org/wiki/United_Nations_Framework_Convention_on_Climate_Change.

(23) United Nations Climate Change. What is the Kyoto protocol? 2020.https://unfccc.int/kyoto_protocol.

(24) United Nations Climate Change. What is the Paris agreement? 2020.https://unfccc.int/process-and-meetings/the-paris-agreement/the-paris-agreement.

(25) Hongbo W. *Global Governance and Global Rules for Development in the Post-2015 ERA*. United Nations; 2014. ISBN 978-92-1-104689-2; eISBN 978-92-1-056769-5.

(26)　UN. *Energy Statistics Pocketbook*. New York: UN; 2019.https://
doi.org/10.18356/1df8f86d-en. https://unstats.un.org/unsd/
energystats/.

(27)　International Energy Agency. Key World Energy Statistics.
2019. https://webstore.iea.org/key-world-energy-statistics-2019.

(28)　IEA. Key world energy statistics. 2019. https://webstore.iea.org/
key-world-energy-statistics-2019.

(29)　Cambridge English Dictionary. Reasonable. 2020. https://
dictionary.cambridge.org/dictionary/english/reasonable.

(30)　United Nations. Sustainable Development Goals. 2020. https://
www.un.org/sustainabledevelopment/. menu=1300

(31)　Roser M, Ritchie H, Ortiz-Ospina E. World population growth.
OurWorldInData.org. 2013.https://ourworldindata.org/
world-population-growth.

(32)　UN-DESA. World population prospects. June 17, 2019. https://
www.un.org/development/desa/publications/world-population-
prospects-2019-highlights.html.

(33)　WorldBank. Nearly half the world lives on less than $5.50 a
day (World Bank Press Release, 2019/044/DEC-GPV). 2018.
https://www.worldbank.org/en/news/press-release/2018/10/17/
nearly-half-the-world-lives-on-less-than-550-a-day.

(34)　Herrling S, Moss T. Scaling power for global prosperity. In: Mair
J, Nee E, Johnson DV, Morgan F, eds. *Stanford Social Innovation
Review: Informing and Inspiring Leaders of Social Change*.
Stanford, CA: Spring; 2019:59–60. https://ssir.org/articles/
entry/scaling_power_for_global_prosperity.

(35)　Canada Energy Regulator. Market snapshot: fuel poverty across
Canada – lower energy efficiency in lower income households.
August 28, 2019. https://www.cer-rec.gc.ca/nrg/ntgrtd/mrkt/
snpsht/2017/08-05flpvrt-eng.html?=undefined&wbdisable=true.

(36)　EIA. One in three U.S. households faces a challenge in meeting
energy needs. 2018. https://www.eia.gov/todayinenergy/detail.
php? id=37072

(37)　Murkowski L, Scott T. Plenty at stake: indicators of American
energy insecurity. 113th Congress, An Energy 20/20 White
Paper. September, 2014. https://www.energy.senate.
gov/public/index.cfm/files/serve?
File_id=075f393e-3789-4ffe-ab76-025976ef4954

(38)　American Council for an Energy-Efficiency Economy.
Understanding energy affordability. 2019. https://www.aceee.
org/sites/default/files/energy-affordability.pdf.

(39) USEIA. 2015 EIA residential energy consumption survey. 2018. https://www.eia.gov/consumption/residential/reports/2015/comparison/.

(40) United States Office of Energy Efficiency & Renewable Energy. LEAD tool. Washington, DC: United States Office of Energy Efficiency & Renewable Energy.https://www.energy.gov/eere/slsc/maps/lead-tool

(41) United States Office of Energy Efficiency & Renewable Energy. Low-income community energy solutions.https://www.energy.gov/eere/slsc/low-income-community-energy-solutions.

(42) US Department of Health & Human Services. Poverty guidelines. January 8, 2020. https://aspe.hhs.gov/poverty-guidelines.

(43) http://worldpopulationreview.com/countries/richest-countries-in-the-world/. Accessed February 20, 2020.

(44) Jollife D, Prydz EB. Estimating international poverty lines from comparable national thresholds (World Bank Group, Policy research working paper 7606). 2016. http://documents.worldbank.org/curated/en/837051468184454513/pdf/WPS7606.pdf.

(45) Hickel J. Could you live on $1.90 a day? That's the international poverty line. *The Guardian*. 2015/2017. https://www.theguardian.com/global-development-professionals-network/2015/nov/01/global-poverty-is-worse-than-you-think-could-you-live-on-190-a-day.

(46) World Bank Group. Access to electricity (% of Population). 2020.https://data.worldbank.org/indicator/EG.ELC.ACCS.ZS.

(47) Worldometer. World population by year. 2020.https://www.worldometers.info/world-population/world-population-by-year/.

(48) IEA. SDG7: data and projections: flagship report. 2019. https://www.iea.org/reports/sdg7-data-and-projections.

(49) United Nations Development Programme. About Sub-Saharan Africa. 2020. https://www.africa.undp.org/content/rba/en/home/regioninfo.html. Accessed March 11, 2020.

(50) United Nations. World population prospects 2019: Highlights (ST/ESA/SER.A/423). 2019.https://population.un.org/wpp/Publications/Files/WPP2019_Highlights.pdf.

(51) Sulligoi G, Vicenzutti A, Menis R. All-electric ship design: from electrical propulsion to integrated electrical and electronic power systems. *IEEE Trans Transp Electrif*. December, 2016;2(4):507-521.

(52)　Mendelsohn M, Lowder T, Canavan B. Utility-scale concentrating solar power and photovoltaics projects: a technology and market overview (NREL technical report, NREL/TP-6A20-51137). April, 2012. https://www.nrel.gov/docs/fy12osti/51137.pdf.

(53)　US-EERE. Renewable energy: utility-scale policies and programs. 2020. https://www.energy.gov/eere/slsc/renewable-energy-utility-scale-policies-and-programs. Accessed February 22, 2020.

(54)　YSG Solar. Utility Scale Solar: What it means and how it works. 2018.https://www.ysgsolar.com/blog/utility-scale-solar-what-it-means-and-how-it-works-ysg-solar.

(55)　Evans P. CBC News, Canada's new power strategy excludes megaprojects. 2011.https://www.cbc.ca/news/canada/canada-s-new-power-strategy-excludes-megaprojects-1.1039996.

(56)　Electricchoice.com. How much electricity on average do homes in your state use? 2020. https://www.electricchoice.com/blog/electricity-on-average-do-homes. Accessed February 22, 2020.

(57)　Shrink. Average household energy use around the world (based on World Energy Council Data). 2014. http://shrinkthatfootprint.com/average-household-electricity-consumption.

(58)　US Environmental Protection Agency. Distributed generation of electricity and its environmental impacts. 2020. https://www.epa.gov/energy/distributed-generation-electricity-and-its-environmental-impacts. Accessed February 22, 2020.

(59)　ABB. Access to electricity: bringing light to rural communities. 2020. https://new.abb.com/sustainability/society/stakeholder-engagement/access-to-electricity. Accessed February 25, 2020.

(60)　Adegoke Y. Africa is facing an electricity crisis – a pay-as-you-go model could solve the problem. *World Economic Forum*. July 2019. https://www.weforum.org/agenda/2019/07/pay-as-you-go-africas-solar-energy/.

(61)　Ivaro L, Wakaba S. *Forget About Power Lines, Pay-As-You-Go is Transforming Africa's Energy Landscape*. London, England: Grantham Institute of Imperial College; March 26, 2018. https://granthaminstitute.com/2018/03/26/forget-about-power-lines-pay-as-you-go-is-transforming-africas-energy-landscape/.

(62)　Eco@Africa.Nigerian company introduces pay-as-you-go solar energy system. 2017. https://www.youtube.com/watch?v=tRqorTVqCXk&t=104s.

(63)   Sarkodia SA. Lessons to be leant from Ghana's excess electricity shambles. *The Conversation.* August 5, 2019. https://theconversation.com/lessons-to-be-learnt-from-ghanas-excess-electricity-shambles-121257.

(64)   Kumi EN. *The Electricity Situation in Ghana: Challenges and Opportunities* (CGD policy paper). Washington, DC: Center for Global Development; 2017. https://www.cgdev.org/publication/electricity-situation-ghana-challenges-and-opportunities.

(65)   Manna S. Ghana's challenges: access to electricity and renewables. January 22, 2018. https://www.aboutenergy.com/en_IT/interviews/ghana-challenges-access-to-electricity-renewables-eng.shtml.

(66)   Watson J, Byrne R, Morgan Jones M, et al. *What are the Major Barriers to Increased us of Modern Energy Services Among the World's Poorest People and are Interventions to Overcome These Effective?* (CEE Review 11-004). UK: Collaboration for Environmental Evidence; 2012. www.environmentalevidence.org/SR11004.html.

(67)   Reader GT. Energy and sustainability: policy, politics and practice. In: Vasel A, Ting D S-K, eds. *EAS 2018.* Cham, Switzerland: SPE; 2019:1-17. https://doi.org/10.1007/978-3-030-00106-6_1.

(68)   North American Electricity Corporation. Definition of "Adequate Level of Reliability", Regional Reliability Plan Guideline. 2017. https://www.nerc.com/docs/pc/Definition-of-ALR-approved-at-Dec-07-OC-PC-mtgs.pdf.

(69)   UN. Our common future: from one earth to one world (Report of the World Commission on Environment and Development). 1987:300. https://sustainabledevelopment.un.org/content/documents/5987our-common-future.pdf.

(70)   Reference 67 Part 1.3 paragraph 27

(71)   Devine D. How long is a generation? Science provides an answer. November 16, 2016. https://isogg.org/wiki/How_long_is_a_generation%3F_Science_provides_an_answer.

(72)   NASA. How long will the Earth remain habitable? https://image.gsfc.nasa.gov/poetry/venus/q79.html. Accessed February 26, 2019.

(73)   Student Energy. Renewable energy. 2020. https://studentenergy.org/source/renewable-energy/.

(74)   WHO. Life expectancy. 2020. https://www.who.int/gho/mortality_burden_disease/life_tables/situation_trends/en/.

(75)   Roser M. Global economic inequality. *OurWorldInData.org*. 2020. https://ourworldindata.org/global-economic-inequality.

(76)   United Nations, Department of Economic and Social Affairs, Population Division. *World Population Prospects 2019, Volume II: Demographic Profiles* (ST/ESA/SER.A/427). New York, NY: United Nations; 2019.

(77)   USEIA. International energy outlook. September 24, 2019. https://www.eia.gov/outlooks/ieo/pdf/ieo2019.pdf.

(78)   Covert T, Greenstone M, Knittel CR. Will we ever stop using fossil fuels? *J Econ Perspectives*. 2016;30(1):117-138.

(79)   Bebbington J, Schneider T, Stevenson L, Fox A. Fossil fuel reserves and resources reporting and unburnable carbon: investigating conflicting accounts. *Crit Perspect Account*. 2020;66:22. doi:10.1016/j.cpa.2019.04.004.

(80)   National Geographic. Non-renewable energy – an encyclopedic entry. 2020. https://www.nationalgeographic.org/encyclopedia/non-renewable-energy/. Accessed February 26, 2020.

(81)   USDOE. 3 reasons why nuclear is clean and sustainable. September 25, 2018. https://www.energy.gov/ne/articles/3-reasons-why-nuclear-clean-and-sustainable.

(82)   USDOE, Office of Nuclear Energy.Nuclear innovation: clean energy future. 2020. https://www.energy.gov/ne/initiatives/nuclear-innovation-clean-energy-future. Accessed February 26, 2020.

(83)   Trabish HK. The unknown costs of a 100% carbon-free future. September 3, 2019. https://www.utilitydive.com/news/the-unknown-costs-of-a-100-carbon-free-future/561639/.

(84)   Loehr GC. The \good\blackout: the Northeast power failure of 9 November 1965 [History]. IEEE Power Energy Mag. May-June 2017;15(3):84–96. doi:10.1109/MPE.2017.2659379

(85)   Nevius D. History of the North American electric reliability corporation. 2020. https://www.nerc.com/AboutNERC/Resource%20Documents/NERCHistoryBook.pdf.

(86)   North America Electric Reliability Cooperation. About NERC. 2020.https://www.nerc.com/AboutNERC/Pages/default.aspx.

(87)   NERC. State of reliability. 2018. https://www.nerc.com/pa/RAPA/PA/Performance%20Analysis%20DL/NERC_2018_SOR_06202018_Final.pdf.

(88)   UNSD. The sustainable development report. 2019. https://unstats.un.org/sdgs/report/2019/The-Sustainable-Development-Goals-Report-2019.pdf.

(89)   UNSD. Tier classification for global SDG indicators. December 11, 2019. https://unstats.un.org/sdgs/files/Tier-Classification-of-SDG-Indicators-11-December-2019-web.pdf.

(90)   UN. Ensure access to affordable, reliable, sustainable and modern energy. https://www.un.org/sustainabledevelopment/energy/.

(91)   UNDP. The energy challenge for achieving the millennium development goals. 2005. https://www.undp.org/content/undp/en/home/librarypage/environment-energy/sustainable_energy/the_energy_challengeforachievingthemillenniumdevelopmentgoals.html.

(92)   Lenardic D. Photovoltaics – historical developments. 2015. http://www.pvresources.com/en/introduction/history.php.

(93)   Davis N. What is the fourth industrial revolution?. *World Economic Forum*. January 19, 2016. https://www.weforum.org/agenda/2016/01/what-is-the-fourth-industrial-revolution/.

(94)   American Physics Society. December 1938: discovery of nuclear fission. *APS News*. December, 2007;16(11):2. https://www.aps.org/publications/apsnews/200712/physicshistory.cfm.

(95)   Husseini F. Riding the renewable wave: tidal energy advantages and disadvantages. *Power Technology*. October 26, 2018. https://www.power-technology.com/features/tidal-energy-advantages-and-disadvantages/.

(96)   USDOE-NOAA. Technical readiness of Ocean Thermal Energy Conversion (OTEC). 2009. https://coast.noaa.gov/data/czm/media/otec_nov09_tech.pdf.

(97)   USDOE-EERE. Renewable hydrocarbon biofuels. https://afdc.energy.gov/fuels/emerging_hydrocarbon.html. Accessed March 1, 2020.

(98)   IEA. Defining energy access: 2019 methodology. 2019. https://www.iea.org/articles/defining-energy-access-2019-methodology.

(99)   European Commission.Towards the super grid for more renewable energy. July 5, 2010. https://ec.europa.eu/jrc/sites/jrcsh/files/20100705_jrc_esof_press_info_grids.pdf.

(100) Grace's Guide to British Industrial Industry. CEGB. 2020. https://www.gracesguide.co.uk/CEGB. Accessed March 3, 2020.

(101) Jones-Albertus B. Confronting the duck curve: how to address over-generation of solar energy. October 12, 2017. https://www.energy.gov/eere/articles/confronting-duck-curve-how-address-over-generation-solar-energy.

(102)   NRCan. Electricity facts. 2019. https://www.nrcan.gc.ca/
        science-data/data-analysis/energy-data-analysis/energy-facts/
        electricity-facts/20068.

(103)   Zochode G. Sell surplus electricity at a discount to Ontario
        businesses rather than exporting at loss, province urged.
        Financial Post. November 21, 2017. https://business.
        financialpost.com/commodities/energy/use-wasted-excess-
        electricity-to-power-economic-growth-engineers-urge-ontario.

(104)   Davis J. Microgrids and neighbourhood power sharing set
        to transform how we use power. *ABC News.* December 4,
        2019. https://www.abc.net.au/news/rural/2019-12-03/
        microgrids-set-to-transform-how-we-use-energy/11756672.

(105)   Desjardins J. Mapped: visualizing the true size of Africa. *Visual
        Capitalist.* February 19, 2020. https://www.visualcapitalist.com/
        map-true-size-of-africa/.

(106)   Patel P. How inexpensive must energy storage be
        for utilities to switch to 100 percent renewables?
        IEEE Spectrum. September 16, 2019. https://
        spectrum.ieee.org/energywise/energy/renewables/
        what-energy-storage-would-have-to-cost-for-a-renewable-grid.

(107)   USEIA. How much ethanol is in gasoline, and how does it affect
        fuel economy? May 14, 2019. https://www.eia.gov/tools/faqs/
        faq.php? id=27&t=10

(108)   IEA. World energy outlook 2019-clean cooking database.
        2019. https://www.iea.org/reports/sdg7-data-and-projections/
        access-to-clean-cooking.

(109)   Karekezi S, Lata K, Coelho ST. Traditional biomass energy.
        A thematic background paper. International Conference for
        Renewable Energies; 2004; Bonn, Germany. 60. https://
        www.ren21.net/Portals/0/documents/irecs/renew2004/
        Traditional%20Biomass%20Energy.pdf.

(110)   Afework B, Hanania J, Stenhouse K, Donev J. Energy education
        - supercritical coal plant. 2018. https://energyeducation.ca/
        encyclopedia/Supercritical_coal_plant.

(111)   WÄRTSILÄ. Combined cycle plant for power generation:
        introduction. https://www.wartsila.com/energy/learn-more/
        technical-comparisons/combined-cycle-plant-for-power-
        generation-introduction. Accessed March 6, 2020.

(112)   Allam R, Martina S, Forrest Bet al. Demonstration of the
        Allam cycle: an update on the development status of a high
        efficiency supercritical carbon dioxide power process employing

full carbon capture. *Energy Procedia*. 2017;114:5948-5966. doi:10.1016/j.egypro.2017.03.1731.

(113)   Power-Technology Cascade. Combined-cycle gas turbine (CCGT) power plant. 2020. https://www.power-technology.com/projects/cascade-combined-cycle-gas-turbine-ccgt-power-plant-alberta/. Accessed March 6, 2020.

(114)   CAN insider. Inside the world's cleanest power plant - In China: coming clean about green. 2018. https://www.youtube.com/watch? v=3dGHLC5YTEA

(115)   USDOE. The history of nuclear energy. DOE/NE Report 0088. 2020. https://www.energy.gov/sites/prod/files/The%20History%20of%20Nuclear%20Energy_0.pdf.

(116)   Nuclear Power. https://en.wikipedia.org/wiki/Nuclear_power#cite_note-bbc17oct-43. Accessed February 29, 2020

(117)   World Nuclear Association. Nuclear power in the world today. 2020. https://www.world-nuclear.org/information-library/current-and-future-generation/nuclear-power-in-the-world-today.aspx.

(118)   International Atomic Energy Agency. Research reactor database (Operational). 2020.https://nucleus.iaea.org/RRDB/RR/ReactorSearch.aspx. Accessed February 29, 2020.

(119)   Hanania J, Jenden J, Stenhouse K, Donev J. Energy education - megawatts electric. 2018. https://energyeducation.ca/encyclopedia/Megawatts_electric. Accessed February 29, 2020.

(120)   US Navy Submarine NR-1. 2020. https://en.wikipedia.org/wiki/American_submarine_NR-1. Accessed February 29, 2020.

(121)   World Nuclear Association. Small nuclear power reactors. 2020. https://www.world-nuclear.org/information-library/nuclear-fuel-cycle/nuclear-power-reactors/small-nuclear-power-reactors.aspx.

(122)   Cho A. *Smaller, safer, cheaper: one company aims to reinvent the nuclear reactor and save a warming plant*. Washington, DC: American Association for the Advancement of Science (AAAS). 2019. https://www.sciencemag.org/news/2019/02/smaller-safer-cheaper-one-company-aims-reinvent-nuclear-reactor-and-save-warming-planet.

(123)   NRCan. Canadian Small Modular Reactor (SMR) roadmap. Summary of key findings. 2019. https://www.nrcan.gc.ca/our-natural-resources/energy-sources-distribution/nuclear-energy-uranium/canadian-small-modular-reactor-roadmap/21183.

(124) McGrath M. Nuclear fusion is a 'question of when, not if'. November 6, 2019. https://www.bbc.com/news/science-environment-50267017.

(125) Devlin H. Nuclear fusion on brink of being realized, say MIT scientists: carbon-free could be 'on the grid in 15 years'. March 9, 2018. https://www.theguardian.com/environment/2018/mar/09/nuclear-fusion-on-brink-of-being-realised-say-mit-scientists.

(126) Manomet Center for Conservation Sciences. Biomass sustainability and carbon policy study (NCI -2010-3). June, 2010. https://www.mass.gov/files/documents/2016/08/qx/manomet-biomass-report-full-hirez.pdf.

(127) Mitchell SR, Harmon ME, O'Connell KEB. Carbon debt and carbon sequestration parity in forest bioenergy production. May 11, 2012. https://onlinelibrary.wiley.com/doi/full/10.1111/j.1757-1707.2012.01173.x

(128) Chen J, Ter-Mikaelian M, Yang H, Colombo SJ. Assessing the greenhouse gas effects of harvested wood products manufactured from managed forests in Canada. *Forestry*. 2018;91:193-205. doi:10.1093/forestry/cpx056.

(129) Holtsmark B. Use of wood fuels from boreal forests will create a biofuel carbon debt with a long payback time. Research Department, Discussions Paper 637, Statistics Norway. November, 2010.

(130) Woodford C. Biofuels. December 13, 2019. https://www.explainthatstuff.com/biofuels.html.

(131) Castanoso J. COP25: EU officials says biomass burning policy to come under critical review. *Mongabay*. December 16, 2019. https://news.mongabay.com/2019/12/cop25-eu-officials-say-biomass-burning-policy-to-come-under-critical-review/.

(132) Heikkinen N. EPA declares biomass plants carbon neutral, amid scientific disagreement. *E&E News*. April 24, 2018. https://www.scientificamerican.com/article/epa-declares-biomass-plants-carbon-neutral-amid-scientific-disagreement/.

(133) IPCC. *Special Report on Renewable Energy Sources and Climate Change Mitigation*. Cambridge University Press; 2012:1088.

(134) BC Government. CleanBC: our nature. Our power. *Our Future*. 2018. https://www2.gov.bc.ca/assets/gov/environment/climate-change/action/cleanbc/cleanbc_.

(135)   Reitig K. The link between contested knowledge, beliefs and learning in European climate governance: from consensus to conflict in reforming biofuels policy. *Policy Stud J.* 2016. doi:10.1111/psj.12169.

(136)   UN.United Nations framework convention on climate change (25pp, FCCC/INFORMAL/84 GE.05-62220 (E) 200705). 1992. https://unfccc.int/resource/docs/convkp/conveng.pdf.

(137)   The Otto Cycle. https://en.wikipedia.org/wiki/Otto_cycle. Accessed February 28, 2020.

(138)   US National Oceanic and Atmospheric Administration. How we measure background CO2 levels at Mauna Loa. 2018. https://www.esrl.noaa.gov/gmd/ccgg/about/co2_measurements.pdf.

(139)   Carslaw KS, Lee LA, Regayre LA, Johnson JS. Climate models are uncertain, but we can do something about it. *Earth and Space Science News.* February 26, 2018. https://eos.org/opinions/climate-models-are-uncertain-but-we-can-do-something-about-it.

(140)   US-NOAA. What's the difference between weather and climate? 2017. https://www.nasa.gov/mission_pages/noaa-n/climate/climate_weather.html.

(141)   US-EPA. Overview of greenhouse gases. https://www.epa.gov/ghgemissions/overview-greenhouse-gases. Accessed February 28, 2020.

(142)   IPCC. Climate change 2013: the physical science basis. Contribution of working group 1 to the fifth assessment report of the IPCC. 2013. https://www.ipcc.ch/site/assets/uploads/2018/02/WG1AR5_Chapter12_FINAL.pdf.

(143)   Lewis N, Crok M. *Oversensitive. The global warming policy foundation report 12.* 2014. https://www.thegwpf.org/content/uploads/2014/03/Oversensitive-download.pdf.

(144)   Lewis N, Curry J. The impact of recent forcing and ocean heat uptake data on estimates of climate sensitivity. *J Climate.* August, 2018;31(15):6051-6071.

(145)   Charney JG, Arakawa A, Baker Jet al. *Carbon Dioxide and Climate: A Scientific Assessment* (Report to the Climate Research Board). Washington, DC: National Research Council, National Academy of Science; 1979:20. https://www.bnl.gov/envsci/schwartz/charney_report1979.pdf.

(146)   USEIA.Annual energy outlook 2020. *Annual energy outlook.* January 29, 2020. https://www.eia.gov/outlooks/aeo/pdf/aeo2020.pdf.

(147)   USEIA. AEO 2020 reference case. 2020. https://www.eia.gov/outlooks/aeo/pdf/AEO2020%20Full%20Infographics.pdf.

(148)   USEIA.The national energy modelling system: an overview 2018. 2019. https://www.eia.gov/outlooks/aeo/nems/overview/pdf/0581(2018).pdf

(149)   BP energy outlook. 2019. https://www.bp.com/content/dam/bp/business-sites/en/global/corporate/pdfs/energy-economics/energy-outlook/bp-energy-outlook-2019.pdf.

(150)   IEA. Global EV outlook 2019. 2019. https://webstore.iea.org/download/direct/2807? fileName=Global_EV_Outlook_2019.pdf.

(151)   UNEP.Climate action summit 2019: a race we can win. September 23, 2019. https://www.unenvironment.org/events/summit/climate-action-summit-2019.

(152)   UN. Report of the secretary-general on the 2019 climate action summit and the way forward in 2020. 2019. https://www.un.org/en/climatechange/assets/pdf/cas_report_11_dec.pdf.

(153)   Appunn K, Wettengel J. Germany's greenhouse gas emissions and climate targets. *Clean Energy Wire*. January 23, 2020. https://www.cleanenergywire.org/factsheets/germanys-greenhouse-gas-emissions-and-climate-targets.

(154)   Hecking H, Hintermayer M, Lencz D, Wagner J. The energy market in n.d. and 2050 – the contribution of gas and heat infrastructure to efficient carbon emission reductions. Final report, Ewi Energy Research and Scenarios. January 2018. https://www.ewi.research-scenarios.de/cms/wp-content/uploads/2017/11/ewi_ERS_Energy_market_n.d._2050_web.pdf.

(155)   Jiang K, He C, Qu C, et al. Are China's nationally determined contributions (NDCs) so bad? *Sci Bull*. 2019;64(6). doi:10.1016/j.scib.2019.01.005.

(156)   Francis D. Canada is a giant Carbon Sink. Why don't we get credit for it?. *Financial Post*. March 6, 2020.

(157)   World Rainforest Movement. The Paris agreement on climate change: promoting tree plantations and reducing forests to tradable carbon stores. 2017. https://wrm.org.uy/articles-from-the-wrm-bulletin/viewpoint/the-paris-agreement-on-climate-change-promoting-tree-plantations-and-reducing-forests-to-tradable-carbon-stores/.

(158)   Levelized cost and levelized avoided cost of new generation resources in the annual energy outlook 2020. February, 2020. https://www.eia.gov/outlooks/aeo/pdf/electricity_generation.pdf.

(159)   Henze V. Battery power's latest plunge in costs threatens
         coal and gas. March 26, 2019. https://about.bnef.com/blog/
         battery-powers-latest-plunge-costs-threatens-coal-gas/.
(160)   Shahan C. Solar costs and wind costs so low they're cheaper
         than *existing* coal and nuclear – Lazard LCOE report.
         CleanTechnica. November 22, 2019. https://cleantechnica.
         com/2019/11/22/solar-costs-wind-costs

CHAPTER 2

Energy Supply and Consumption in Senegal

Bertrand Tchanche

Department of Physics, University Alioune Diop of Bambey, Bambey, Senegal

Abstract

Senegal is located in West Africa with a population close to 16 million inhabitants unequally distributed on a land of 196,722 km^2 area. In the 2000s, a national energy information system (known as SIE-Sénégal) aiming at monitoring and forecasting the energy demand and the efficient planning of the energy infrastructure was put in place in the Ministry of Oil and Energies. A lot of data were recorded of which some are analyzed and presented here for a better understanding of the energy system of Senegal. In the period 2000–2013, the energy demand has been increasing reaching 3.72 Mtoe in 2013. The demand is covered by imported fossil fuels and traditional biomass. The energy consumption has been increasing in the same period from 1.69 Mtoe in 2000 up to 2.56 Mtoe in 2013. The energy pattern shows a lion's share for the residential sector followed by the transport and industrial sectors. In the residential sector, firewood is the main fuel, and electricity is deemed marginal. The transport sector is dominated by the road subsector where diesel oil represents 81% of the energy use. In the industrial sector, more than 80% of energy used is from fossil origin and the share of coal is becoming significant.

Keywords: energy, Senegal, transport, industry, residential

1.0   Introduction

Senegal is a West African country of about 16 million inhabitants, of which around 23% live in the capital city, Dakar—see Table 1 [1]. It has a land area of about 196,722 km² and an average population density of 73 inhabitants per kilometer. The population growth rate is 2.87% per year: from 4.3 in 1971, the number of inhabitants reached 15 million in 2017 (see Fig. 1). The climate is hot tropical, the country being located in the Sahel region, between Sahara desert and African savanna. The economy of the country relies on agricultural products (groundnuts, cotton, fish, fruits, and vegetables), minerals (gold, phosphates), and tourism. The gross domestic product (GDP) per capita based on

Table 1: Population Estimates and Distribution (*Source*: Data Compiled From ANSD. [1])

	Region	Population	Share (%)
1	Dakar	3,529,300	23.13
2	Diourbel	1,692,967	11.10
3	Tambacounda	783,777	5.14
4	Saint-Louis	1,009,170	6,61
5	Thies	1,995,037	13.08
6	Fatick	813,542	5.33
7	Kaffrine	655,121	4.29
8	Kaolack	1,086,464	7.12
9	Kedougou	172,482	1.13
10	Kolda	748,451	4.91
11	Louga	976,885	6.40
12	Matam	654,981	4.29
13	Sedhiou	517,016	3.39
14	Ziguinchor	621,168	4.07

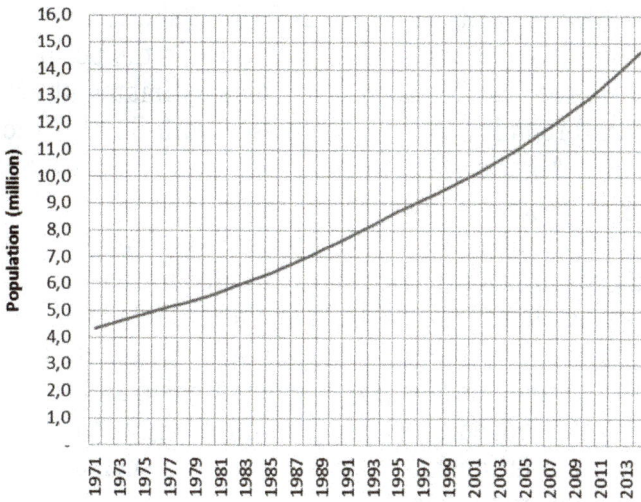

Fig. 1: Demography Dynamics (*Source*: Data Compiled by the Author From National Statistics and Demography Agency. [2]).

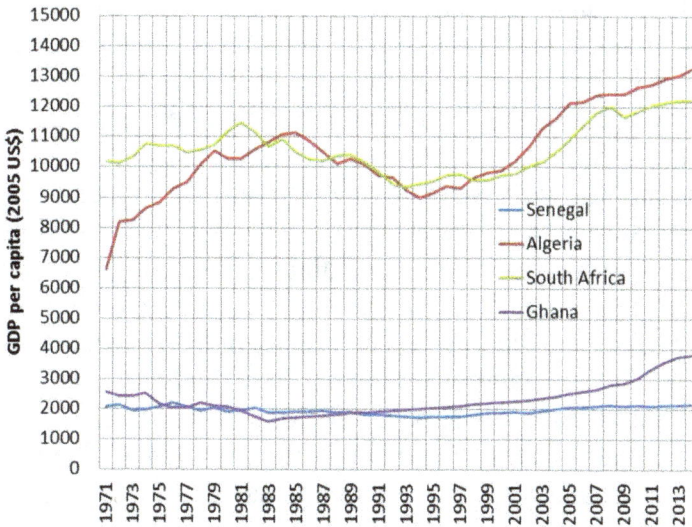

Fig. 2: GDP Comparison of Senegal With Other African Countries (*Source*: Data Compiled From IEA (International Energy Agency) [2]).

power purchasing parity (PPP) in 2005 US$ shows that on a capita basis it has not changed significantly in the period 1971–2014—see Fig. 2, and in 2014, was estimated at around US$ 2181 and remains low in comparison with South Africa (US$ 12,197), Algeria (US$ 13,267), and Ghana (US$ 3,815) [2].

Energy is progressively being recognized as one of the pillars of the modern economies; not only natural resources, capital and human resources are necessary but also infrastructures and energy. Energy should be available and affordable for sustainable growth. The energy system of Senegal as many others in sub-Saharan Africa has not been extensively studied and lot of research is still needed to understand the energy consumption pattern and its relationship with other sectors such as transport, agriculture, education, health, etc.

In rural areas, traditional agriculture is the main activity. It uses animal and human energy, and very little commercial energy. Energy is needed at any step of farming, from land preparation till the harvesting, storage, and commercialization. But lack of appropriate financial support mechanism that would have enabled farmers to modernize their farms through the acquisition of modern equipment and improved storage facilities does not allow them to increase their productivity and their revenue. Energy is important for education in many ways. Available and affordable electricity enhance significantly the quality of education; not only that modern pedagogic material can be used by educators but students should have good environment for studies at home and access to online materials. The quality of health is associated with clean energy availability. Electricity is critical for health centers, and its lack as well as its interruptions cause deaths and serious damages. The quality of energy used is also important. Use of traditional biomass in households as well as in some activities in the services sector and that of dirty fuels in the transport sector causes respiratory problems. The role of energy for productive activities and industrial companies is being progressively accepted in sub-Saharan Africa. The energy crisis experienced by many countries in the 2000s caused reduction of their GDP as companies where not able to satisfy their clients' demands and extend their activities. Many countries became less attractive as investors did not want to develop business in a risky environment. Utilization of backup systems (e.g., diesel engines) as a continuous energy supply system affects negatively the economy, by fueling the inflation as additional cost is passed on to consumers. The energy crisis of the 2000s brought to the light the necessity to study the energy flows and transformation within the country. In the Ministry of Oil and Energies, a specific office was set in order to collect and analyze statistical data related to the energy sector. As part of the policy of the country, a number of agencies dedicated to energy issues were put in place: [3] AEME (Agence pour l'Economie et la Maitrise de l'Energie) created in 2011 is in charge of energy efficiency, ASER (Agence Sénégalaise d'Electrification Rurale), conducts rural electrification projects, and ANER (Agence Nationale pour les Energies Renouvelables) created in 2013 promotes renewable energy resources. Survey shows lack of study regarding the energy system of Senegal, and the present paper is produced to provide an analysis and offers possible debate on the energy policy of the country.

2.0 Energy Resources

The country possesses untapped and diverse energy resources. Biomass is found almost everywhere in the country, but the Southern and Eastern parts of the country have good potential (Tambacounda, Ziguinchor, Kolda, Kaolack). Biomass constitutes main source of energy especially for domestic practices. Firewood is harvested from farms and forests, dried and part of it transformed into charcoal, and shipped into urban areas. It is the principal energy source in rural areas. However, this resource although very important and strategic has not been managed sustainably and half of the forest has been lost since 1960 [4]. Senegal has huge solar energy potential, although unevenly distributed with northern part presenting highest potential. The average global solar radiation is about 2,000 kWh/m^2/year [5]. Studies have been carried out to determine the potential but lot still needed for total resource assessment and characterization, and a solar map is not yet available. Attempts to develop photovoltaic systems have been made but the installed capacity remains low, far behind the technical potential. Solar domestic hot water systems have not received a lot of attention, and only few installations are found in Dakar. Investigations on wind potential in Senegal showed exploitable resources in western part of the country. According to Bilal et al. [6], coastal areas from Saint Louis to Dakar have monthly mean wind speed between 3 and 6 m/s (height 12 m) and could be suitable for commercial wind farms. Several sites with good potential have been identified and projects are under development: Taiba (125 MW), Mboro (50 MW), Leona (50 MW), and Saint Louis (50 MW) [7]. Hydroelectricity is exploited, and the Manantali hydropower plant has a capacity of 200 MW but other possibilities exist on Senegal and Gambia Rivers [8]. OMVS (Organisation de la Mise en Valeur du Fleuve Sénégal) and OMVG (Organisation de la Mise en Valeur du Fleuve Gambie) have identified suitable sites with total technical potential above 2 GW. Offshore wind, tidal, and wave energy could be developed on the maritime area but studies have to be done to prove feasibility. Oil and gas fields have been discovered recently in the country and exploration missions are to confirm the potential. Meanwhile, the Senegalese government is looking for strategies for sustainable exploitation.

3.0 Electricity Generation

Senegal has not yet reached the universal access to energy, and the challenge is huge. Notwithstanding the increasing electrification rate observed since 2000, the global electrification rate was about 61.7% in 2017 [9]. While the urban electrification rate is close to 90%, in rural areas it is still too low, less than 20% as of 2013 (see Fig. 3). The electricity generation has two origins: fossil fuels (thermal power plants) and renewable energy sources (wind, solar, and hydro). Fuel oil, diesel-, and gas-driven plants make up the thermal capacity. Hydropower, wind power, and solar photovoltaic plants complement the installed capacity. Minor part of this generation capacity comes from agricultural waste-based power plants.

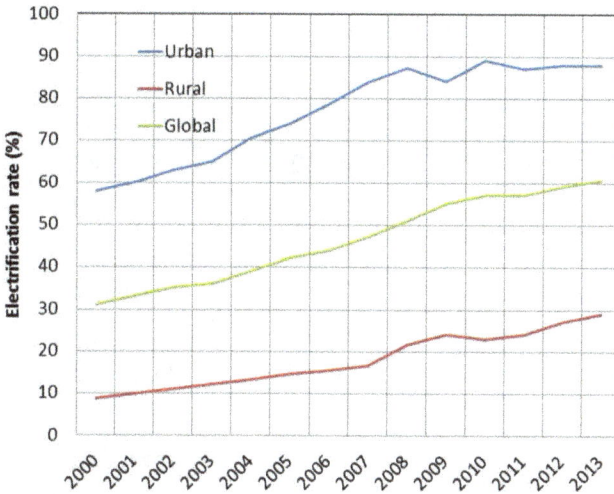

Fig. 3: The Evolution of the Electrification Rate(Adapted by the Author From SIE Reports).

Electricity is generated by a set of actors including Senelec, which is the public power company (PPC), independent power producers (IPP), and self-power producers (Cim-sahel Energy, CSS, Sonacos, Sococim, and ICS). Senelec is a monopolistic company in charge of production, transport, and distribution. It is a state-owned company authorized to sign contracts with independent producers to purchase their production and sale it to consumers. Senelec owns and operates majority of thermal plants in the country but its operating capacity is insufficient and in order to cover the increasing demand has signed contracts with private producers like Aggreko and APR Energy LLC. Aggreko holds four stations (Cap des biches 1 and 2, Boutoute and Diass) and APR Energy LLC [10] two (Tambacounda and Kanoune). Table 2 presents a nonexhaustive list of electrical power plants operated or rented by Senelec.

The number of IPPs has been increasing over the years as a result of the change in the energy policy of the country, which encourages private investments. Accordingly, the legislative framework was modified and a regulatory authority established. Table 3 lists a number of thermal power plants operated by independent producers. These plants rely on heavy fuel oil (HFO),

Table 2: Nonexhaustive List of Power Plants Operated or Rented by Senelec

Producers	Plants	Capacity (MW)	Fuel Type	Operation
Senelec	CVI/601	16.45	Heavy fuel oil	2006
	CVI/602	16.45	Heavy fuel oil	2006
	CVI/603	16.45	Heavy fuel oil	2006
	CVI/604	16.45	Heavy fuel oil	2006
	CIV/401	21	Heavy fuel oil	1990
	CIV/402	21	Heavy fuel oil	1990
	CIV/403	23	Heavy fuel oil	1997
	CIV/404	15	Heavy fuel oil	2003
	CIV/405	15	Heavy fuel oil	2003
	CIII/301	27.5	Heavy fuel oil	1966
	CIII/302	30	Heavy fuel oil	1975
	CIII/303	30	Heavy fuel oil	1978
	CIII/TAG2	18	Kerosene	1984
	CIII/TAG3	20	Gasoil	1995
	CII/TAG4	30	Gasoil	1999
Rented Plants	APR Energy, Kanoune	67.5	Heavy fuel oil	2007

Producers	Plants	Capacity (MW)	Fuel Type	Operation
	APR Energy, Tambacounda	6		
	Sococim	14		
	Aggreko/cap des biches 1	50	Heavy fuel oil	
	Aggreko/cap des biches 2			
Imports	Mauritania	20		

diesel oil (DO), and natural gas. There has been a debate on whether the Sendou power plant originally designed as a coal power plant should be turned into a gas power plant. The proposal was made by the government who intend to reduce the burden of imported fossil fuels, by using natural that will be produced in the country in the upcoming years with the combined advantage of more energy independence and cleaner, affordable, and reliable electricity.

Table 3: Nonexhaustive List of Thermal Power Plants Developed and Operated by IPPs (Independent Power Producer) *Source:* CRSE [11]

Producers	Plants	Fuel type	Capacity (MW)	Contract (years)	Operation
Contour Global	Cap des biches (Dakar)	DO and HFO	89.2	20	2000
Kanoune Power	Kanoune (Dakar)	HFO	67.5	15	2008
Tobene Power SA	Tobene (Dakar)	HFO	20	117.8	2016
Charbon Sendou CES	Bargny Minam (Dakar)	Coal/ gas	25	125	2018
Malicounda Power SAS	Malicounda (Thiès)	Dual fuel	120	20	Ongoing

In recent years, the State has put in place strategies to exploit renewable energy resources for electricity generation, and a number of projects have been successfully implemented, while some are still in the pipeline waiting for funding. An emphasis has been put on large-scale grid-connected plants and a favorable legislative framework was set to allow promoters of renewable energy-based power plants to sell their production to Senelec. An ambitious plan for 30% of renewables in the electricity mix by 2018 was adopted by the government and since 2016 several plants were put online, few of them being listed in Table 4

Table 4: Nonexhaustive List of Solar Photovoltaic and Wind Projects (From IPPs) *Source:* CRSE [11]

Name/Source	Location	Capacity (MWc)	Contract (years)	Operation
Senergy II/solar	Bokhol/saint Louis	20	25	2016
Senergy PV SA/ solar	Santhiou-Mékhé/ Thiès	29.5	25	2017
Energy Resources Senegal SA/ solar	Kahone/ Kaolack	20	25	2018
Ten Merina/ solar	Merina (Dakhar)	29.5	25	2017
INNOVENT SA/ solar	Sakal/Louga	20	25	2018
EPC of KfW/ solar	Dias/dakar	23		2019
Groupement Solaria Kima/ solar	Malicounda/ Mbour	22	25	2016
ENGI Meridien (Scaling solar)	Kahone (Kaolack)	35		Ongoing

Name/Source	Location	Capacity (MWc)	Contract (years)	Operation
ENGI Meridien (Scaling solar)	Kael (Touba)	25		Ongoing
Taiba Ndiaye SA /Wind farm	Taiba Ndiaye (Thiès)	151.80		Ongoing

Senegal is a member country of the Economic Community of West African States (ECOWAS), which in 1999 created the West African Power Pool (WAPP). Member States recognized that the vast energy resources within the ECOWAS region could be harnessed for the mutual benefit of all Member States [12]. In 2006, the WAPP was established with the mission to promote and develop infrastructure, for power generation and transmission, as well as, to assure the coordination of electric power exchanges between ECOWAS countries. The WAPP has the status of a specialized institution of ECOWAS. The WAPP works in collaboration with other entities of ECOWAS such as the ECOWAS Centre for Renewable Energy and Energy Efficiency (ECREEE), the ECOWAS Regional Electricity Regulatory Authority (ERERA), as well as the West African Gas Pipeline (WAGP) Company. As part of the WAPP, several projects have been implemented as listed in Table 5 [11]. The 200 MW Manantali hydropower plant

Table 5: Nonexhaustive List of Projects Developed in the Framework of WAPP (West African Power Pool), Developed by IPPs *Source:* CRSE [11]

Plants	Project framework	Technology	Capacity (MW)	Operation
Manantali	OMVS	Hydro	66	2002
Felou	OMVS	Hydro	15	2013
Kaleta	OMVG	Hydro	100	Ongoing
Gouina	OMVS	Hydro	35	Ongoing
Kokoutamba	OMVS	Hydro	70	Ongoing
Sambagalou	OMVG	Hydro	128	Ongoing

is located in Mali supplies 32% of power generated to Senegal, and 15 MW out of 60 MW from Felou hydropower plant is made available for Senegal.

As of 2017, the total installed electrical capacity was 1,021 MW (84 MW offgrid), while the available capacity was 836 MW distributed as follows:[11] thermal (89%), hydro (6%), and solar (5%). Part of this capacity is out of operation due to aging equipment and poor maintenance. Majority of capacity is made up of thermal power plants and the remainder based on renewables—hydro, solar, and agricultural residues and around 20 MW are imported from Mauritania. It is worth mentioning that frequent power cuts have made diesel engines very popular. Despite their increasing number, records fail to establish neither their number nor the energy produced. In 2018, availability and capacity factors were 87% and 65%, respectively [9]. In 2017, total electricity generated amounted to 3.92 TWh of which 55% were from Senelec and the remainder from private producers[13]. Grid-connected plants generated 3.73 TWh, and off-grid generation is thus marginal.

The maximum electrical power peak is observed in October, and has been increasing over the years as can be seen in Fig. 4—from 246 MW in 2000 it has more than doubled, reaching 533 MW in 2015 [14], and 642 MW in 2018, one of the highest in the ECOWAS region after Nigeria 16.66 GW, Ghana 2.71 GW, and Cote d'Ivoire 1.38 GW. The peak demand to on-grid capacity ratio is 0.53, with a total on-grid capacity estimated at 1201 MW in 2018 [9].

The electrical network shown in Fig. 5 is made up of two main parts, an interconnected grid and a nonconnected grid (regional minigrids and isolated systems). The interconnected grid (RI) receives a capacity of 803 MW. This part of the network is centered on western and northwestern parts of the country and supplies Dakar, Thies, Louga, Diourbel, Saint Louis, Matam, Kaolack, and Fatick. Plants connected to RI are Senelec (444 MW), GTI and Kanoune Power (119 MW), Manantali and Felou (75 MW), APR and Aggreko (145 MW), and imports (20 MW). The nonconnected grid is made up of two regional minigrids (Tambacounda and Ziguinchor) and 23 isolated plants (disseminated in Kaolack, Tambacounda, Kolda, and Ziguinchor), all based on DO with a total capacity of 83 MW. Electricity produced is transported through two levels high voltage (HV) lines, of 90 kV and 225 kV.

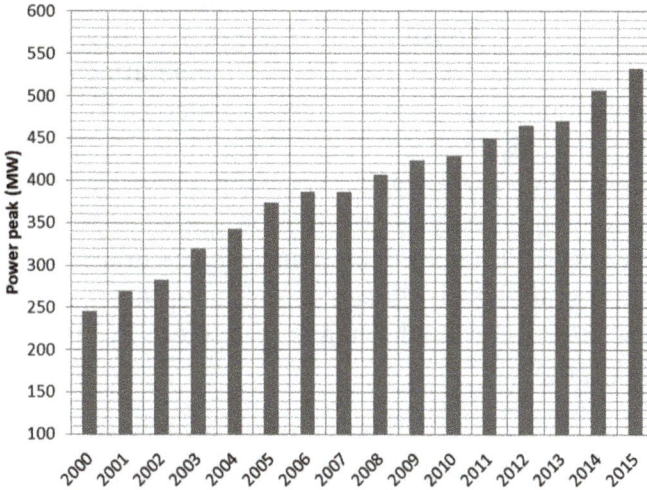

Fig. 4: The Peak Power Evolution (Produced by the Author From the Senelec Annual Report).

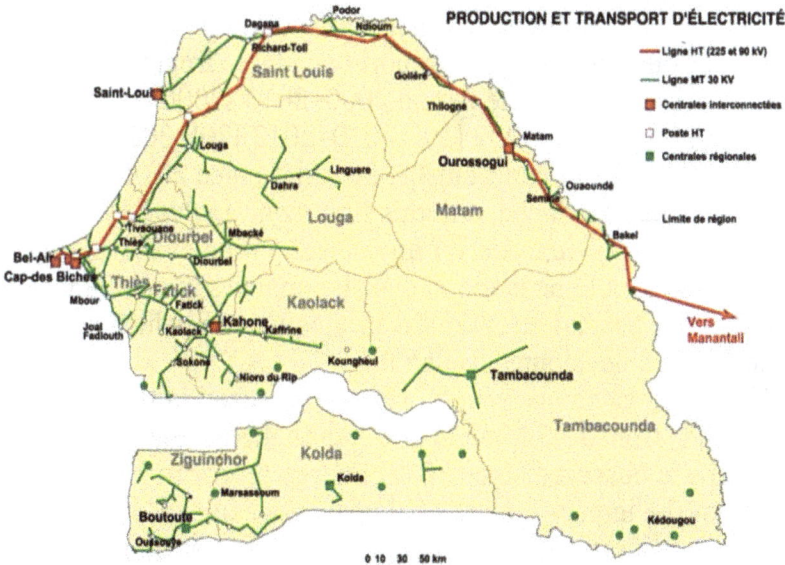

Fig. 5: Mapping of the Senegalese Electrical Network (Source: Senelec Annual Report)

The average end-user tariff (EUT) for domestic use is around US\$ 0.24, and is a very sensitive matter in the country. Any attempt to increase the price could lead to social tensions. Therefore, any government wishing to last should integrate this reality, and should effort to reduce or keep constant the electricity unit price. Fossil fuels represent a real burden to Senelec—it represents about 86.2% of the operating costs [9], and any reduction in oil prices is always welcome. Senelec does not operate in profitable mode, and is kept alive by the subsidies of the government. The electricity demand grows by 6% [15] per year and could hardly be covered for a number of reasons: operation of the PPC under debts, massive use of imported fossils fuels, nonintegrated grid, old equipment, nonprofitable and old small-scale plants, low maintenance, poor electrical network management, etc.

4.0   The Energy Supply

The total primary energy supply (TPES) is the balance between the inflow energy to the country, outflow, generation, and stock variation. Data utilized in this section and the next ($5) were extracted from various reports including SIE (Système d'Informations Energétiques) reports available from the Ministry of Oil and Energies for the period 2000–2013 (2000–2004 [16], 2005 [17], 2006 [18], 2009 [19], 2010–2012 [20], 2013 [21]). Author would like to point out the fact that since 2014 no SIE report has been produced, and this limits our analysis. Real and updated data are important for any investigation of the energy system. Call should be made to decision makers in sub-Saharan Africa to give special attention to data collection, which will further help in development planning.

The TPES as can be seen in Fig. 6, increased in the period 2000–2011, from 2.45 Mtoe in 2000 up to 4.62 Mtoe in 2011 then, a decrease is observed since 2012. Energy production represents 87.20% of the imports and the balance between exports and imports shows the dependency of the country over imports and raises concern over energy security. From this, the enthusiasm of the government is understandable after the gas discovery of recent years that could reduce the burden of fuel imports.

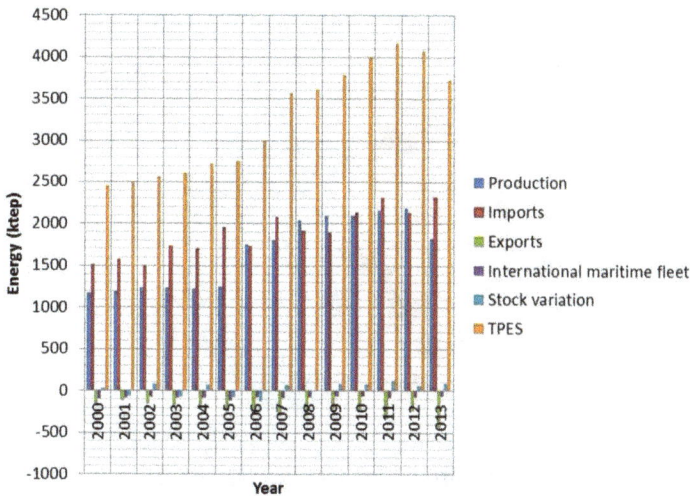

Fig. 6: Total Primary Energy Supply(Adapted by the Author From SIE Reports)

The share of fossil fuels in TPES is almost equal to that of renewable energy sources, and that of waste heat is deemed marginal–see Fig. 7. Figures 8 and 9 show that in 2006, fossil fuels imports dropped, and were partly compensated by renewables. Figure 8 shows a sharp increase of renewables from 2005. Biomass and specifically firewood dominates (94.35% in 2013). Although vast solar resource is found, it does not appear in the supply. Senegal imports all fossil fuels used in the country and has an oil refinery plant operated by the state-owned company SAR (Société Africaine de Raffinage). Figure 9 shows that fossil fuels demand has been increasing in the period 2000–2013 and was around 1.9 Mtoe in 2013. It is observed that the proportion of oil products is gaining momentum, from 2006. The sole refinery in the country is no longer able to satisfy the increasing demand, due to lack of investment in revamping the plant. A debate about the management of this company surged some years ago, and experts question why such important and strategic company is not well managed and lacks investments. Part of oil products imported is of low quality, and its combustion generates pollutants that cause serious health issues. The growth in coal demand observed is due to the increasing number of cement factories (Sococim, Dangote Cement, and Ciments du Sahel).

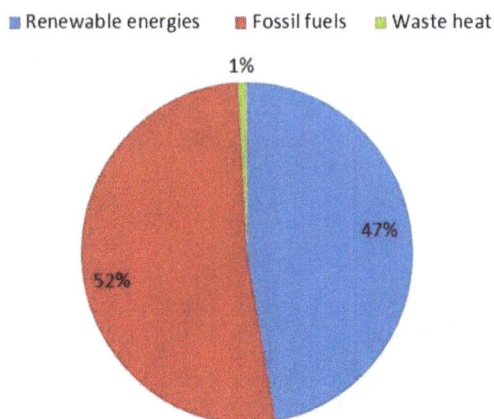

Fig. 7: Share of Energy Types in TPES for the Year 2013 (Adapted by the Author From SIE Reports)

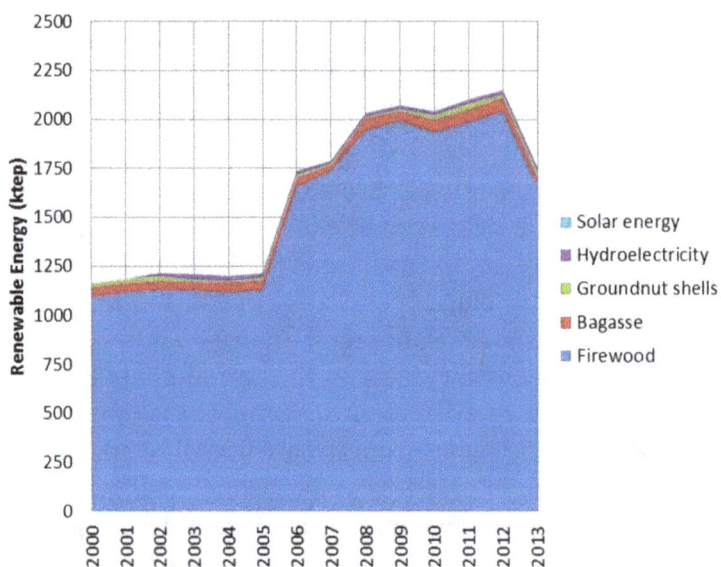

Fig. 8: Dynamics of Renewable Energy Supply in the Period 2000–2013 (Adapted by the Author From SIE Reports).

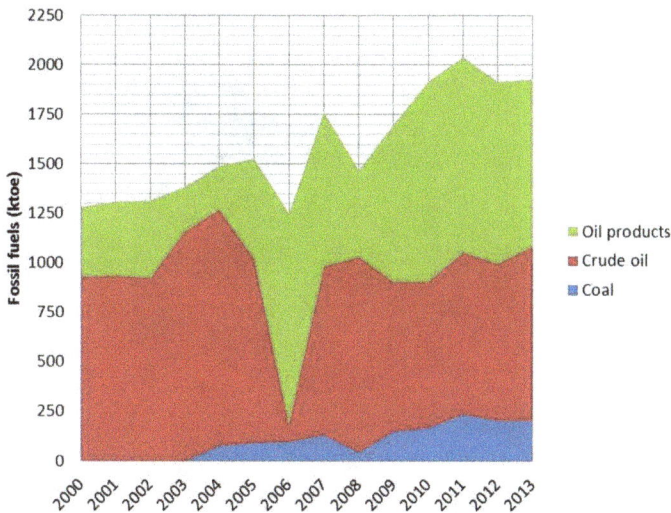

Fig. 9: Fossil Fuels Dynamics in the Period 2000–2013 (Adapted by the Author From SIE Reports)

5.0　The Energy Consumption

5.1　Total Final Energy Consumption

Final energy refers to the energy utilized by end users: households, commercial buildings, factories, etc. Figure 10 displays the evolution of the total final energy consumption in the period 2000–2013. The total final energy consumption has been increasing from 1.69 Mtoe up to 2.56 Mtoe in 2013. The residential sector has the largest share in 2013, representing 48% of total final energy consumed and followed by transport 30% and industry 16%—see Fig. 11. While the increase observed was due to the residential sector, industrial and transport sectors didn't experience significant change. Services and agriculture use very little quantity of energy. Agriculture is not modernized, and still based on traditional practices, with very little mechanization. Nevertheless, the energy consumed in the services may be underestimated or partly merged with the residential sector.

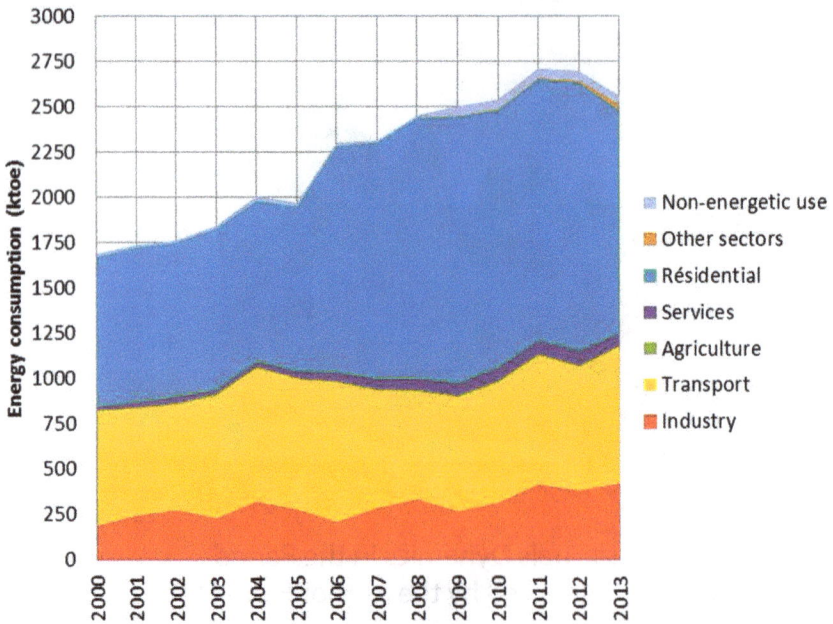

Fig. 10: The Energy Consumption Dynamics in Different Sectors in the Period 2000–2013 (Adapted by the Author From SIE Reports)

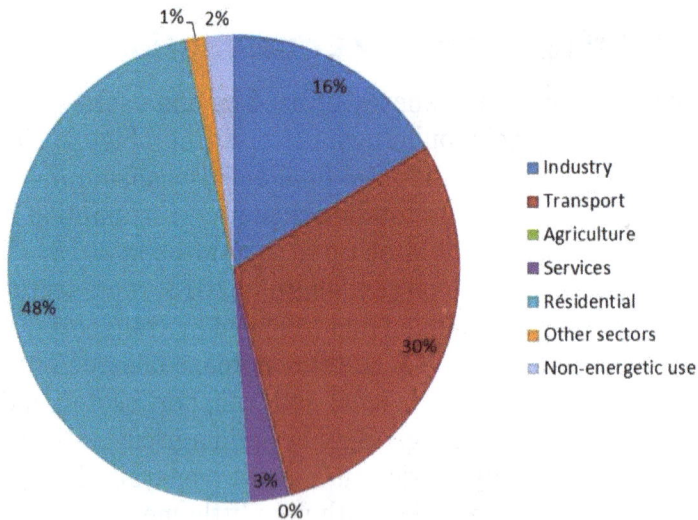

Fig. 11: Energy Consumption Distribution of Sectors (2013) (Adapted by the Author From SIE Reports)

Rigorous evaluation of the energy consumption of the services would mean separating activities bound with domestic activities and those done out of households—this looks difficult in the context of Senegal.

5.2   *Total Energy Consumption in the Transport Sector*

Transport sector is well understood if the colonial period is taken into account. First, at the beginning of that period (1800s till 1960s), maritime transport on rivers was most used to travel from coastal areas to inner parts of the country. Rail took over when resources were to be transferred from West Africa to Europe (and mainly France). The railway was the most developed transport system till the independence of the country and the first line Dakar–Saint Louis was inaugurated in 1885. It was part of the strategy designed by France to connect several countries, including Mauritania, Niger, Mali, and Senegal and for the transportation of goods and passengers, in what was called AOF (Afrique Occidentale Française). After the independence of Senegal in 1963, the first free government decided to change this strategy and to focus on national transport, and gave more emphasis on roads. And since that period rail was progressively abandoned in the national policy, and received less investments. This led to the reduction of traffics, the number of passengers reduced and some lines disappeared. The total of rail lines in the country is about 906 km [1]. Two companies operate in this sector, Transrail SA for international traffic on the line Dakar–Bamako and Petit Train de Banlieue SA for the transportation of passengers between Dakar and its suburbs. A new line is under construction between Diamniadio and Dakar, to facilitate the transport of passengers between Dakar and the new airport Blaise Ndiagne. Today, with the development of road transportation facilities with the advantages associated such as flexibility, conviviality, and rapidity, the railways has less importance, and do represent marginal part of transport activities. Data related to energy consumption in this subsector were scarce, and, therefore, not considered in this study.

One of the priorities of the first free government led by Leopold S. Senghor was to build infrastructure in order to connect various parts of the country, and foster the feeling of nationalism. It is seen from records and archives that local entrepreneurs imported first cars early in the 1950s. 1970s saw the installation of assembly lines from many European companies like Berliet and Mercedes. But those manufacturers didn't last long in the country, because of the smaller size of the local market and lack of appropriate government policy. Today, about 80% of vehicles of the national car fleet are second-hand imported from Europe, America, and Asia. In 2010, according to the National Statistics Bureau, 59% of cars bear diesel engines and 38% use gasoline [1]. Most of these cars are concentrated in Dakar (+73%), about 365,000 cars [22]. National car fleet is characterized by old vehicles, with average age being 20 years and light-duty vehicles represent 85% of the total [23].

Senegal has 14 airports, of which main ones are located in Dakar, Saint-Louis, Cap Skiring, Tambacounda, and Ziguinchor. Air transportation with national companies was seen after the independence as a sign of sovereignty and prosperity and any independent government in sub-Saharan Africa was proud to own one. In Senegal, there have been attempts to develop national aviation and past experience has been with Air Afrique, Air Senegal International, or Senegal Airlines. Air Senegal International was created in 2000 after an agreement between Royal air Maroc (RAM) and the state of Senegal. The state held 49% of auctions of the capital. Partners diverged about the strategy of the company, and this led to the end of the activities in 2008. Senegal Airlines was founded in 2009, after an agreement between the government and local investors, and started operations in 2011 but the mismanagement of the company led to a debt (over US$ 168 million), which forced to the end in 2016 [22].

Senegal has four sea ports and Dakar (Port Autonome de Dakar—PAD) is the most important. Most of trade is done through maritime transport. Traffic at the PAD is close to 12.2 million tons per year [22]. The Ziguinchor sea port is the second port after Dakar. This is justified by the fact that the main way of reaching Casamance region is by sea. Kaolack sea port is specialized in oil, oilcake, and salt exports.

Energy mix in the transport sector shows a domination of road transport. It is seen from Fig. 12 that the share of energy consumed on roads has been increasing over the years, while energy consumed by maritime was almost constant and that consumed by air transport was dramatically reduced since in 2007. As mentioned earlier, rail was not taken into account because of lack of data. The picture presented is in line with the development in the transport sector. Air transport in general is not well developed in sub-Saharan African countries, and most companies struggle and face strong competition from foreign companies. Mismanagement, high ticket prices, poor maintenance, and inefficiency are some difficulties they face [24]. Maritime transport is limited to coastal areas and no clear development policy is seen. Joola catastrophe brought to the light the mismanagement on the main line Dakar-Ziguinchor. The unique ship that linked Casamance and Dakar sunk one night and thousands of lives were lost. Gasoil is the main fuel on roads due to its price, lower than that of gasoline. Its consumption has been increasing as seen in Fig. 13, while that of gasoline remained almost constant over the years. Kerosene has been disappearing

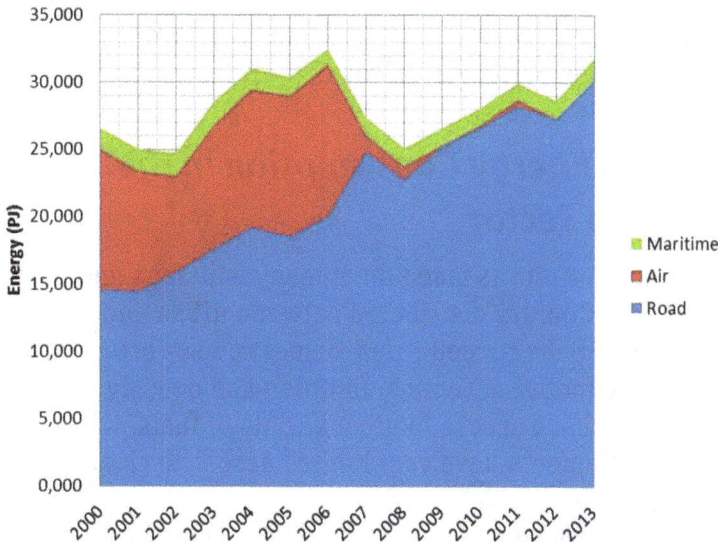

Fig. 12: Dynamics of Energy Consumed in Different Modes (Adapted by the Author From SIE Reports)

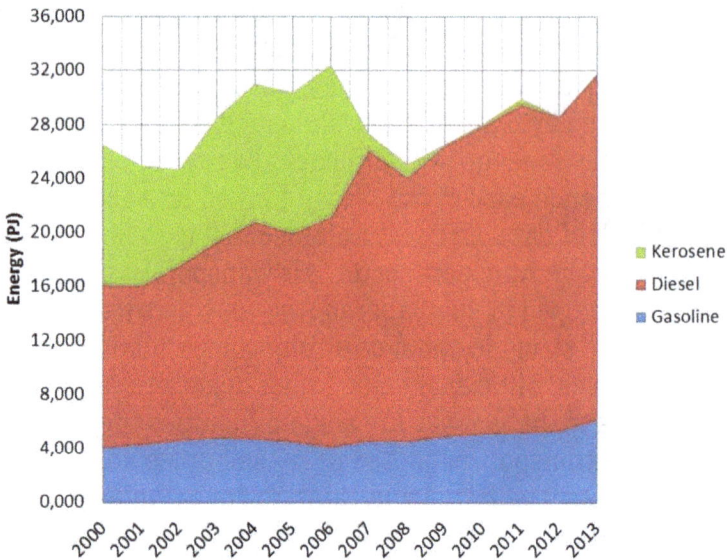

Fig. 13: Energy Fuels Dynamics in the Transportation Sector in the Period 2000–2013 (Adapted by the Author From SIE Reports)

from statistical data since 2007, and in 2013 diesel accounted for 81% and gasoline 19%. For the same year, 95% of energy consumed in transports was on roads [25].

5.3 Total Energy Consumption in the Residential Sector

The residential sector is made up of many buildings and houses in all parts of the country distributed between urban and rural areas. As in most sub-Saharan countries, domestic uses are combined with small productive activities. In urban and densely populated areas dwellers have access to great variety of fuels, while those in remote settlements have very limited access to clean and commercial fuels.

The energy consumption in the residential sector has been increasing, from 826 up to 1477 ktoe in the investigated period as seen in Fig. 14, even though a decrease is observed in 2013.

Again, it would have been interesting to have data covering the period after 2013, but these are not available. However, the energy pattern is dominated by traditional biomass (firewood and charcoal), and the share has been more important since 2005. In 2013, firewood accounted for 53%, charcoal 30%, liquefied petroleum gas (LPG) 9%, electricity 8%, and kerosene less than 1% (see Fig. 15). Biomass is thus the dominant fuel, accounting for 83% and is mainly used for cooking. Cooking activities bear particular cultural and social significance, and are usually carried out by women who are exposed to air pollution. The popularity of biomass is justified by its availability and cost. It is the most affordable fuel. Biomass is found in vast forests located in the southern and eastern parts of the country. Trees are cut, dried, and shipped in urban areas. This has led to deforestation of some parts of the country. In an effort to alleviate negative effects of unsustainable use of biomass, the government has set some programs aiming at developing improved cooking stoves (an example is the PROGEDE [Projet de Gestion Durable et Participative des Energies Traditionnelles et de Substitution] project) and spreading small-scale biogas plants. LPG has progressively become a paramount choice in urban households. It has many advantages over traditional biomass, as it is easily carried, is available and produces less pollutants. The government has encouraged its utilization by providing subsidies. Now, it represents close to 9% of energy consumed in the residential sector—see Fig. 15. Electricity is the less used type of energy, and justification could be the lowest electrification rate as majority of population leave in rural areas without access to electricity. Low-income and high electricity cost are likely to limit the use of electricity consumption. Only necessary appliances would be acquired by households for lighting, leisure, communication, and few productive activities. Air conditioning, washing machine, and water heater are just out of possibilities for most citizens.

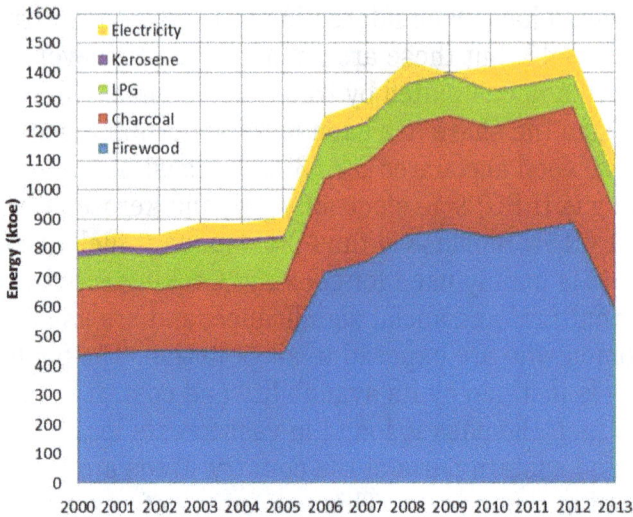

Fig. 14: Energy Fuels Dynamics in the Residential Sector in the Period 2000–2013 (Adapted by the Author From SIE Reports)

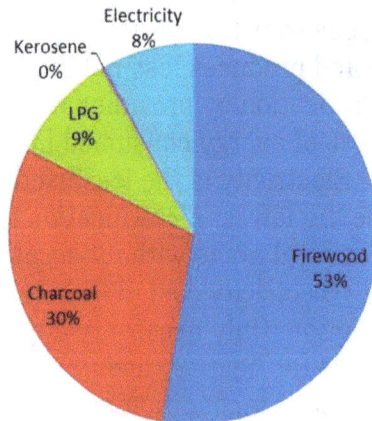

Fig. 15: Energy Fuels Distribution in the Residential Sector in 2013 (Adapted by the Author From SIE Reports)

5.4   *Total Energy Consumption in the Industrial Sector*

The industrial sector is a crucial sector in the economy, and determines the level of the development of the country. Many studies have pointed out the fact that the lack of reliable and affordable energy is impeding the development of African nations. Thus, it becomes a priority if a country want to attract foreign investments. Here, the energy consumption in the industrial sector is analyzed.

The energy consumption in the industrial sector of Senegal has been increasing in the period 2000–2013, from 185.1 up to 417 ktoe (see Fig. 16). The industrial energy mix has been dominated by fossil fuels, made up of fuel oil and diesel oil till 2003, and then the fraction of coal became progressively large. By 2013, coal accounted for 52% followed by electricity 18%, fuel oil 13%, and the remainder covered by LPG, bagasse, diesel, oil and groundnut shell (see Fig. 17). Statistical data obtained from different SIE reports do not allow to clearly categorizing types of industries

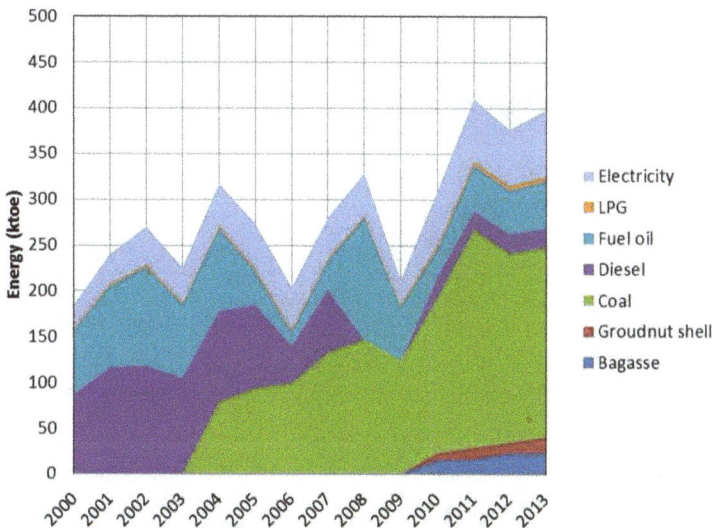

Fig. 16: Energy Fuels Dynamics in the Industrial Sector in the Period 2000–2013 (Adapted by the Author From SIE Reports)

before 2003. Nevertheless, in the period 2000–2013, the share of energy consumed by nonmetallic mineral companies has been increasing and exceeds by far that of other categories (see Fig. 18). In 2013, 67% of the energy consumed in the industrial sector was by nonmetallic mineral companies, 9% by agro-food, 8% by chemical and petrochemicals, and building materials, textiles, iron, and steel, and not identified industries cover the remainder.

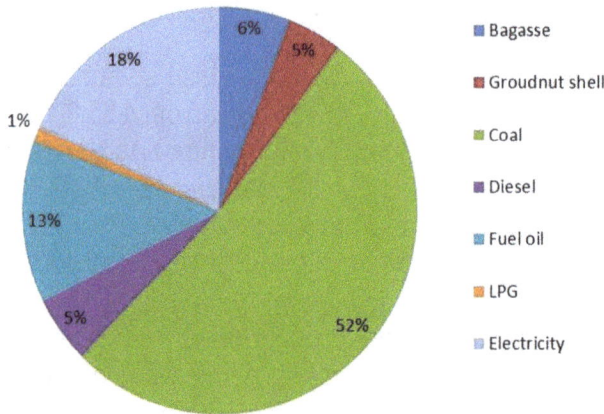

Fig. 17: Energy Fuels Consumption Distribution in the Industrial Sector in 2013 (Adapted by the Author From SIE Reports)

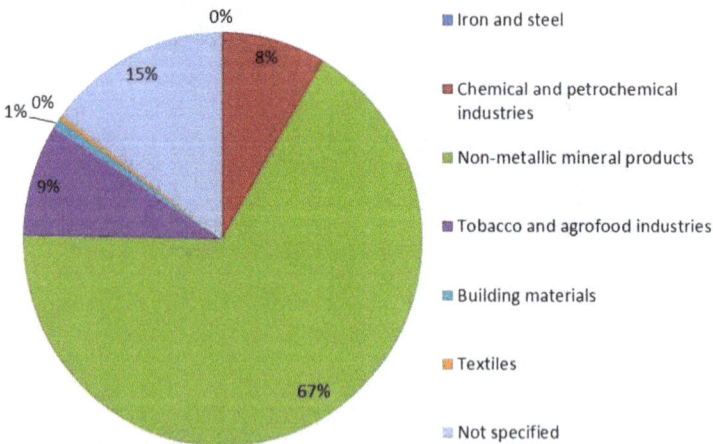

Fig. 18: Energy Consumption Distribution Industry Type in 2013 (Adapted by the Author From SIE Reports)

6.0   Conclusion

Analyzing the energy system of a country is an important task, as it allows better understanding of the dynamics of the energy flow and impacts on other sectors including the economy. Here has been assessed the energy system of Senegal. Main findings are as follows: (1) electricity generation and transport system rely on fossil fuels imports and brings to the light the issue of energy independence and the sustainability of the economy; (2) massive use of biomass in the residential sector with the advantage of increasing the energy security of the country; (3) an industrial sector heavily relying on fossil fuels imports, and (4) very low utilization of energy in the agricultural sector. Recent developments show that the country has gas reserves, and the exploitation is due to start in the upcoming years and will certainly increase the energy security of the country. Efforts should be done to preserve biomass resources, and some projects go in that direction. Renewable electricity is being considered as a priority, and few projects have been successfully implemented and the trend is expected to continue.

References

(1)   Agence Nationale de la statistique et de la démographie [Internet]. 2020. http://www.ansd.sn/.

(2)   IEA. *CO2 Emissions From Fuel Combustion*. Paris, France: International Energy Agency; 2016.

(3)   Ministère du Pétrole et des Energies. 2020. http://www.energie.gouv.sn/les-agences/.

(4)   Gaudreau K, Gibson RB. Sustainability assessment of the agricultural and energy systems of Senegal. *Energies*. 2015;8(5):3503-3528.

(5)   Dia D, Fall CS, Ndour A, Sakho-Jimbira MS. *Le Sénégal face à la crise énergétique mondiale: enjeux de l'émergence de la filière des biocarburants*. Bame: ISRA Bame; 2009:52.

(6)   Bilal BO, Ndongo M, Kebe CMF, Sambou V, Ndiaye PA. Feasibility study of wind energy potential for electricity

generation in the Northwestern Coast of Senegal. *Energy Procedia*. 2013;36:1119-1129.

(7) Thiam A. *Rapport de l'étude de marché du solaire thermique: production d'eau chaude et séchage des produits agricoles*. IRENA: Dakar; 2015:74.

(8) IRENA. *Senegal Renewables Readiness Assessment*. Abou Dhabi, UAE: International Renewable Energy Agency; 2012:76.

(9) AfDB, ERERA.*Comparative Analysis of Electricity Tariffs in ECOWAS Member Countries*. Abidjan, Cote d'Ivoire; 2019:135.

(10) Thiam D-R, Benders RMJ, Moll HC. Modeling the transition towards a sustainable energy production in developing nations. *Appl Energy*. 2012;94:98-108.

(11) Comité de régulation du secteur de l'électrocité. Producteurs indépendants. 2020. http://www.crse.sn/index.php/producteurs-independants.

(12) ECOWAS. *2020 - 2023 WAPP Business Plan*. Abuja, Nigeria; 2019:107.

(13) CRSE. *Rapport Annuel 2017*. Dakar, Sénégal: Comité de régulation du secteur de l'électricité; 2017:80.

(14) Senelec. *Rapport Annuel 2015*. Dakar, Sénégal: Société Nationale d'Electricité; 2015.

(15) Tchanche B. Analyse du système énergétique du Sénégal. *Rev Energ Renouv*. 2018;21(1):73-88.

(16) MEDER. *Rapport SIE-Sénégal/2005*. Dakar, Sénégal; 2005.

(17) MEDER. *Rapport SIE-Sénégal/2006*. Dakar, Sénégal; 2006.

(18) MEDER. *Rapport SIE-Senegal/2007*. Dakar, Sénégal; 2007:56.

(19) MEDER. *Rapport SIE-Sénégal/2010*. Dakar, Sénégal; 2010.

(20) MEDER. *Rapport SIE-Sénégal/2013*. Dakar, Sénégal; 2013:64.

(21) MEDER. *Rapport SIE-Sénégal/2014*. Dakar, Sénégal; 2014:55.

(22) Tchanche B. Energy consumption analysis of the transportation sector of Senegal. *AIMS Energy*. 2017;5(6):912-929.

(23) Tchanche B. Exergy analysis of the transportation sector of Senegal. *Afr J Environ Sci Technol*. 2019;13(8):310-316.

(24) Tchanche B. Concevoir des systèmes de transport durables en Afrique. *Liaison Energ Francoph*. 2018;108:31-33.

(25) Tchanche B, Diaw I. Analyse énergétique du secteur des transports du Sénégal. In: *Proc. 1ère Conférence Ouest Africaine des Energies Renouvelables*, 28 Juin - 02 Juillet. Saint-Louis, Sénégal; 2017.

CHAPTER 3

Hydrogen as an Energy Vector: Present and Future

Isabel Amez Arenillas, Marcelo F. Ortega, Javier García Torrent, Bernardo Llamas Moya
Universidad Politécnica de Madrid, Madrid, Spain

Abstract

The energy sector is undergoing a transformation, from fossil fuels to renewable sources. The expansion of these latest technologies is usually based on sources such as wind or photovoltaic. Therefore, the need arises to search for energy storage processes, to decouple production from demand. Moreover, intermittent renewable energies are unpredictable. Hydrogen technology emerges as one of the most promising option to store huge amount of energy (necessary to manage electricity network). Hydrogen is not a renewable resource, although it can be produced by renewable energy sources like solar energy, wind energy, and hydropower. Therefore, this chapter aims to review the four squares of hydrogen implementation in order to evaluate the state of the art, resulting in a proposal for a storage system able to connect the main cornerstones of hydrogen. Moreover, the proposed system tries to overcome the current barriers that delimit the use of renewable energies.

Keywords: Hydrogen, hydrogen storage, hydrogen production, electrolysis, hydrogen mixtures.

1.0 Introduction

Hydrogen is the lightest and simplest atom of all the periodic system elements. Under normal conditions of pressure and temperature, it is a nontoxic, colorless, and odorless gas, characterized by its highly flammable nature.

Despite hydrogen being the most abundant element in the universe, unfortunately, it does not directly constitute an available combustible fuel, as it is an energy vector and not an energy source. The potential of hydrogen as an energy vector has aroused great interest in recent years due to its versatility, since it is possible to use it in many of the sectors, most commonly used in present society.

The energy sector is undergoing a transformation, from fossil fuels to renewable sources. The expansion of these latest technologies are usually based on sources such as wind or photovoltaic. Therefore, the need arises to search for energy storage processes, in order to decouple production from demand. Moreover, intermittent renewable energies are unpredictable. Hydrogen technology emerges as one of the most promising option to store huge amount of energy (necessary to manage electricity network). Hydrogen is not a renewable resource, although it can be produced by renewable energy sources like solar energy, wind energy, and hydropower. Considering these sources, it can be called green Hydrogen Production as Hydrogen is produced without greenhouse gas (GHG) emissions [1]. Nevertheless, currently, about 95% of the produced hydrogen stems from fossil fuel.

The use of this combustible presents interesting advantages, such as its high-energy density based on the mass rate [2], its high availability and its properties, like stability and noncorrosion, and fundamentally its "clean combustion" with air. The reaction in Equation 1 only produces nitrogen oxides (NOx), depending on H_2/air rate [3].

$$H_2 + O_2 \leftrightarrow H_2O + \text{heat} \tag{1}$$

However, it is also important to consider its disadvantages, like the complexity of its transportation and the high cost of hydrogen, its low-energy density based on volume rate and the need to use other energy sources to produce it. In fact, some authors warn in their latest studies about the hydrogen leaks, which could seriously affect the ozone layer [4].

The advantages of using hydrogen make it one of the biggest research areas nowadays. The necessity to overcome barriers such as its transportation, safe usage, and adaptation of infrastructures—among others— has aroused a great interest lately.

The understanding of hydrogen as an energy vector for the future starts with the study of four cornerstones that will drive the implementation, production, use, storage, and safety of hydrogen. Figure 1 summarizes the "Hydrogen Square" as proposed by Furat Furat [5].

One of the four cornerstones, the hydrogen storage, is linked to another of the most studied areas, the energy storage. The balance of renewable energy production goes through energy storage, since its intermittence generates an excess of electricity, specifically in off-peak hours. If we are capable to store those

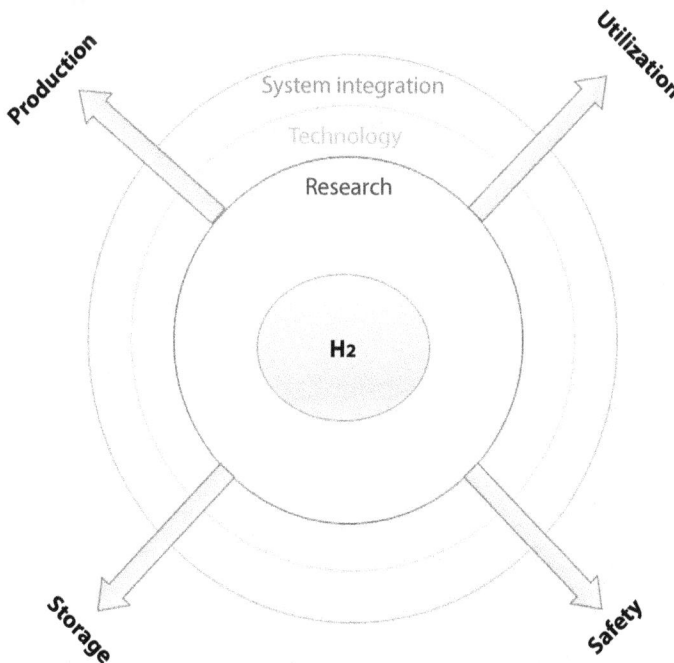

Fig. 1: Hydrogen Square

excesses through other energy vectors, like hydrogen, the imbalance of the current energy system could be solved.

Therefore, this chapter aims to review the four squares of hydrogen implementation in order to evaluate the state of the art, resulting in a proposal for a storage system able to connect the main cornerstones of hydrogen. Moreover, the proposed system tries to overcome the current barriers that delimit the use of renewable energies.

2.0   Hydrogen Production

As already mentioned, approximately 95% of hydrogen is produced by fossil fuels. Depending on the way to produce hydrogen, it can be classified as gray, blue, or green [6]. Gray hydrogen is the one produced by conventional methods without pollutant removal systems. Nevertheless, when conventional methods incorporate systems to capture and store CO_2 and other pollutants, hydrogen is considered blue. Finally, hydrogen originated from renewable energies is considered as green hydrogen. Figure 2 presents the main hydrogen production methods, classifying them depending on their origin (green, blue, and gray hydrogen).

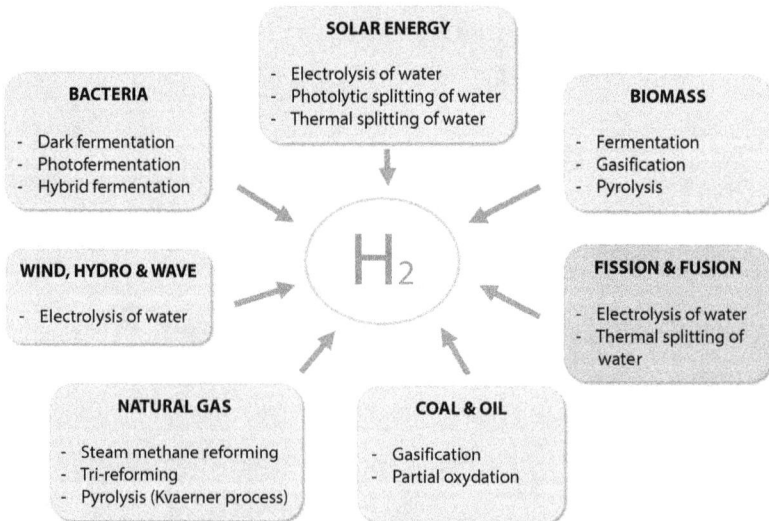

Fig. 2: Classification of the Hydrogen Production Processes Depending on Their Origin

In addition, the main processes to obtain hydrogen are described below.

2.1 Gray and Blue Hydrogen

2.1.1 Steam Reforming

The steam reforming (SR) process consists of producing hydrogen as of hydrocarbons, normally from natural gas. This process has two steps to carry out the separation of the carbon contained in the CH_4 molecule. The first one takes place in the primary reformer, whereas HC the gases and the deionized water steam are combined with a rate of $H_2O:C$, about 2.5–4, being approximately 20%–25% methane concentration. The mixture enters inside the reactor approximately at 200°C and 35 bar, reacting with the catalyst at 800°C and converting about 50% of CH_4.

After the primary reactor, the output steam contains around 10%–13% of CH_4. This steam enters in the secondary reactor, wherein the reaction take place at more temperature, between 900°C and 1000°C, which decreases the CH_4 to about 1% approximately.

The water-gas shifts and the pressure swing adsorption (PSA) are the last stages to purify hydrogen, obtaining as a product stream with 99.9% of hydrogen. The global efficiency of the process is between 70% and 90%.

$$CH_4 + H_2O \rightarrow 3H_2 + CO \tag{2}$$

2.1.2 Partial Oxidation of Fossil Fuels with Oxygen Deficient (POx)

This method consists of heating natural gas or hydrocarbon inside a reformer, in the presence of an atmosphere with oxygen deficient. Reactions 3 and 4 produce hydrogen steam that needs to be purified. The temperature of the water steam supplies the needed energy to produce the reaction.

$$CH_4 + 0.50_2 \rightarrow CO + H_2 \tag{3}$$

$$CH_4 + O_2 \rightarrow CO_2 + 2H_2 \tag{4}$$

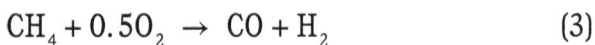

Partial oxidation is expensive, as the method requires pure oxygen, although it has clear advantages, like its low sensitivity to fuel variations, compactness, and quick response time. This fact makes POx a useful method for mobile applications [7].

Partial oxidation can be catalytic (CPOx) and thermal partial oxidation (TPOx). Thermal POx requires high temperatures and is characterized by its flexibility to work with a wide range of fuels like methane or heavy fuel, for instance. On the other hand, catalytic oxidation can be an alternative for the SR process, since SR is extremely endothermic. Moreover, TPOx works at lower temperatures than POx, reducing the rate of temperatures by approximately 590K [8].

2.1.3 *Autothermal Reforming*

Autothermal reforming (ATR) combines the partial oxidation technology and the steam reforming technology in a single reaction. First, natural gas is oxidized into syngas, and secondly, the carbon monoxide reacts with the water steam to produce CO_2 and H_2 that will be captured in an amine absorption process. The main advantage of this combined process is the use of the excess heat produced in POx to fulfil the heat demand of SR.

Catalysts are essential in the ATR process, being a key point in the effectiveness and associated costs.

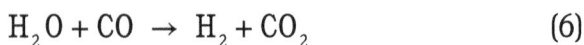

$$CH_4 + H_2O \rightarrow 3H_2 + CO \tag{5}$$

$$H_2O + CO \rightarrow H_2 + CO_2 \tag{6}$$

2.1.4 Thermo-Electrolysis

The increase in temperature under the operating conditions allows obtaining hydrogen with high purity and efficiency. Thermo-electrolysis works with temperatures around 700°C–1000°C. Additionally, the reaction kinetics is favored at high temperatures, resulting in a decrease of electric losses.

2.1.5 Thermolysis

The thermolysis process split water directly through heat, breaking down water molecules into oxygen and hydrogen. The temperatures are normally around 2500°C, being the most important point of the material resistance. Nevertheless, one of the biggest problems of thermolysis is gas separation. The process does not allow the separation of hydrogen and oxygen until the temperature decreases. Consequently, the flammable mixture composed of hydrogen and oxygen has to be stable during the cooling time of the gases.

2.2 Green Energy

2.2.1 Hydrogen Production from Biomass

Hydrogen can be obtained by solid fuels like biomass. There are fundamentally two procedures: biomass gasification and pyrolysis.

The biomass gasification involves the incomplete burning of biomass at temperatures between 700°C and 1200°C, which mainly produces a combustible gas composed of hydrogen, methane, and carbon dioxide.

Nevertheless, the pyrolysis is not a combustion reaction, as this method is a thermal decomposition of the biomass in the absence of oxygen at 500°C. In this case, the products are solid carbon, liquids, and a gas mixture composed of carbon monoxide, carbon dioxide, hydrogen, and light hydrocarbons.

2.2.2 *Biological Production of Hydrogen*

The main advantage of the biological production of hydrogen is the atmospheric working conditions. Atmospheric conditions allow reducing costs in comparison with chemical processes. Biophotolysis, photo-fermentation, and dark fermentation are the most promising methods.

Biophotolysis: Some bacteria like cyanobacteria, or some plants like algae, can produce hydrogen by means of solar light, water, and enzymes [9]. Biophotolysis can be direct or indirect. The direct process uses microalgae like green algae and cyanobacteria. The chemical reaction is described below.

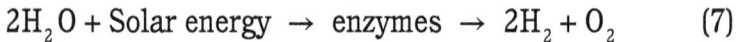

$$2H_2O + \text{Solar energy} \rightarrow \text{enzymes} \rightarrow 2H_2 + O_2 \qquad (7)$$

The indirect process has two steps. The first one is based on photosynthesis, where it results in glucose. In the second step, water reacts with glucose using the light energy to generate hydrogen and CO_2. Blue-green algae (cyanobacteria) is considered in this process [10].

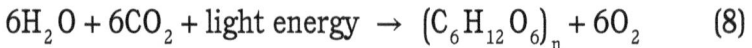

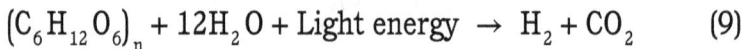

$$6H_2O + 6CO_2 + \text{light energy} \rightarrow \left(C_6H_{12}O_6\right)_n + 6O_2 \qquad (8)$$

$$\left(C_6H_{12}O_6\right)_n + 12H_2O + \text{Light energy} \rightarrow H_2 + CO_2 \qquad (9)$$

Both processes are currently under research due to its low efficiency.

Photo-fermentation: Photosynthetic bacteria such as *Rhodopseudomonas palustris* [11] or *Rhodobacter spheroids* [12], are able to produce hydrogen under certain conditions. The reaction occurs in the presence of water, light energy, and organic acids with the help of a nitrogenase enzyme. The reaction is described below.

$$C_6H_{12}O_6 + 12H_2O + \text{light energy} \rightarrow 12H_2 + 6\,CO_2 \qquad (10)$$

The most interesting aspect of this biological process is the possibility to use biomass wastes to produce hydrogen. Nevertheless, some disadvantages like low efficiencies and high costs due to the oxygen deficient conditions, make this process not available nowadays, as it happened with biophotolysis.

Dark fermentation: This method is the only biological process of hydrogen production carried out without light [13]. The reaction occurs in the dark in presence of anaerobic bacteria and microalgae under thermophilic conditions. The products of this process are mostly carbon dioxide and hydrogen, combined with other gases such as methane. The methane production reduces the efficiency of the process. Therefore, it is necessary to carry out a pretreatment of inoculum at 70°C to kill methanogens and reducing the methane production.

Some new researchers propose the combination between dark fermentation and photo-fermentation to ferment biological wastes. The combination of both processes can increase the efficiency of hydrogen conversion [14].

2.2.3 Electrolysis

An electric current passing through water produces dissociation between hydrogen and oxygen, components of the H_2O water molecule. Hydrogen is collected at the cathode (negatively charged pole) and oxygen at the anode. The process is much more expensive than steam reforming, but it produces high-purity hydrogen. This hydrogen is used in the electronics, pharmaceutical, or food industry.

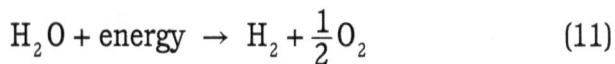

$$H_2O + energy \rightarrow H_2 + \frac{1}{2}O_2 \qquad (11)$$

Currently, the use of electrolysers to produce hydrogen through the dissociation of water is an implemented system in the industry.

Table 1. compare principal electrolyzes. The main electrolysers are described below.

Table 1: Comparison of the Main Electrolysers [15,16]

	Alkaline	Proton Exchange	Solid Oxide Electrolysis
Temperature	20°C–80°C	20°C–200°C	500°C–1,000°C
Efficiency	59%–82%	65%–82%	up to 100%
Operating range	20%–100%	5%–100%	–
Applicability	Commercial	Close to commercialization	Laboratory scale
Advantages	1. Low capital cost 2. Stability 3. Mature technology	1. Compact systems 2. Fast response 3. Fast start-up	1. Kinetic and thermodynamic improvement 2. Lower energy demands 3. Low capital cost
Dis advantages	1. Corrosive electrolyte 2. Hydrogen/ oxygen permeation 3. Slow dynamics 4. Hydrogen purification	1. High cost 2. Use of noble metals 3. Low OH- conductivity in polymeric membarnes	1. Mechanically unstable ceramic membranes 2. Safety problems 3. Sealing leakages
Challenges	1. Durability 2. Reliability 3. Efficiency	1. Electrolyte improvement 2. Reduce noble metal utilization 3. Reach a PEM efficiency around 94%	1. Microstructural properties in electrodes 2. Carbon deposition 3. Passivation

3.0   Alkaline Water Electrolysis

Alkaline technology is the most consolidated electrolyser worldwide [15]. This technology fundamentally consists of two electrodes in the presence of an aqueous alkaline solution, normally KOH and/or NaOH. The electrolyte concentration is about 20%–30% and the electrode materials can be made by nickel. After the water dissociation, the produced gas has a hydrogen concentration of around 99% purity. However, the final gas contains an alkaline fog that must be removed before the purification process. A desorption process removes the alkaline fog. The diaphragm of alkaline water electrolysis (AWE) is made of asbestos, a porous material in which the oxygen/hydrogen that penetrates has an associated explosion risk. By means of pressure control, it is possible to reduce the risk [15].

4.0   Proton Exchange Membrane Water Electrolysis

Proton exchange membrane (PEM) electrolysers are based on proton exchange technology by means of membranes. The membrane conducts the protons and, therefore, it replaces the asbestos conductors. PEM technology is more environmentally friendly since it does not produce alkaline fog. Its compact design made it one of the most promising technologies for water dissociation. PEM technology is useful when the energy supply varies. That is the case of renewable energy. Nevertheless, PEM technology is more expensive than alkaline and nowadays AWE is more used in large-scale plants [16].

Some researchers consider PEM electrolysis the most promising technique to produce hydrogen from water [17]. The reason is essentially the high pure efficient hydrogen production from renewable energy sources and the fact that this process emits only oxygen as a by-product. PEM technique does not emit carbon emissions [16].

The proton exchange membranes have numerous advantages such as high proton conductivity, high-pressure operations, lower thickness, and lower gas permeability. In the future, characteristics like compact design, high efficiency, small footprint, fast response high current density, operates under lower temperatures and produces high pure hydrogen will make PEM electrolysers an important tool for green energy production and for balancing of renewable energy production.

For PEM technology, the future research direction should be the improvement of PEM electrolysers adaptability to renewable energies and develop more cost-effective electrolysers [16].

5.0   Solid Oxide Electrolysis

Solid oxide electrolysis (SOE) technology has aroused great interest due to the production of hydrogen with ultra-pure hydrogen and greater efficiency. The operation conditions are high pressure and high temperatures, about 500°C–850°C utilizing steam. SOE process conventionally uses pure oxygen conductors, normally from nickel/yttria-stabilized zirconia [16]. Current studies try to develop some innovative proton-conducting ceramic materials for SOE. These materials are the key to obtaining the highest efficiency and superior ionic conductivity than oxygen conductors. The operation temperature of ceramic conductors will be around 500°C–700°C. The main characteristic of SOE technology is the higher operating temperature, which makes it advantageous compared to low-temperature electrolysis. However, SOE electrolysers need to overcome serious issues for its commercialization, like its lower stability and the material degradation.

Therefore, the development of stable and active cathodes for SOE is one of the key point for the future. The cathode improvement manly depends on microstructures and the material intrinsic properties [18].

One of the most promising aspects of SOEs technology is CO_2/H_2O coelectrolysis. This technology is still immature owing to the poor carbon dioxide tolerance of the ion/proton conduction electrode. Therefore, the future of SOE cell (SOEC) technology lay on the improvement of ion conduction SOECs. A typical SOE comprises a "sandwich" as assemble containing an oxygen ion/proton conduction electrolyte, an anode and a cathode. Reactions 12–15 control the coelectrolysis process of water and carbon dioxide.

(a) Ion-conducting electrode
   Cathode:

$$CO_2 \, (g) + 2e^- \rightarrow CO(g) + O^{2-} \tag{12}$$

$$H_2O(g) + 2e^- \rightarrow H_2 \, (g) + O^{2-} \tag{13}$$

   Anode:

$$O^{2-} \rightarrow O_2 \, (g) + 4e^-$$

(b) Proton-conducting electrode
   Cathode:

$$CO_2 \, (g) + 4H^+ + 4e^- \rightarrow CO + H_2 + H_2O \tag{14}$$

   Anode:

$$2H_2O(g) \rightarrow 4H^+ + O_2 + 4e^- \tag{15}$$

While AWE is a consolidated technology owing to the durability and reliability, PEM is getting more and importance, specifically talking about green energy production from renewable energies. Both technologies have a potential role in the near future to integrate wind and solar energy to the current electricity network. However, PEM allows working with a wide range of electricity demand, making it more flexible [19].

By contrast, SOEC is an immature technology that is still under research. Nevertheless, its potential for some of the most important uses in the medium-term future has boosted great interest and, nowadays, is one of the most important research areas in the electrolysis field [20].

From an economical point of view, PEM and AWE technology have led to impressive cost reductions, and the estimations foresee a greater reduction in the next years [21]. SOEC technology is an expensive technology but its importance is increasing, consequently, it may be expected a reduction in the cost for next decades as well.

5.1 Photoelectrolysis or Photoelectrochemical Process

The photoelectrolysis is an electrolysis in which the electrical current is originated by solar energy, more specifically, by solar radiation. Solar radiation is transformed into chemical energy through hydrogen. In photoelectrochemical process (PEC), the sunlight interacts with semiconductors, which split water into H_2 and O_2. There are numerous photoelectrodes such as Fe_2O_3, n-GaN, n-GaAs, WO_3, or TiO_2, among others, that offer reasonable efficiencies but TiO_2 is currently the most promising.

5.2 Ion Exchange Membranes or Solid Polymer Electrolyte

Water is split at the anode into O_2, protons, and electrons; the protons pass through the membrane, while the electrons are transferred through an external electric circuit to the cathode. At the cathode, the protons and electrons combine to produce H_2. These steps can be illustrated according to the following chemical reactions:

Anode:

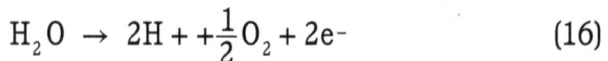

$$H_2O \rightarrow 2H + +\frac{1}{2}O_2 + 2e- \qquad (16)$$

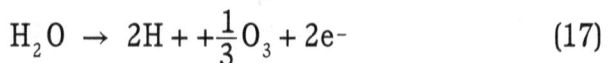

$$H_2O \rightarrow 2H + +\frac{1}{3}O_3 + 2e- \qquad (17)$$

Cathode:

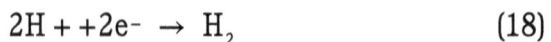

$$2H + +2e- \rightarrow H_2 \qquad (18)$$

6.0 Hydrogen Utilization

The demand for hydrogen is expected to increase by 4%–5% per year over the next 5 years [22]. Figure 3 shows some of the future uses of hydrogen with the highest utilization potential.

Conventional hydrogen utilization is focused on refineries and the chemical industry. However, the medium-to-large-term future foresees new potential uses of hydrogen as Fig. 3 describe. The most important ones are described below.

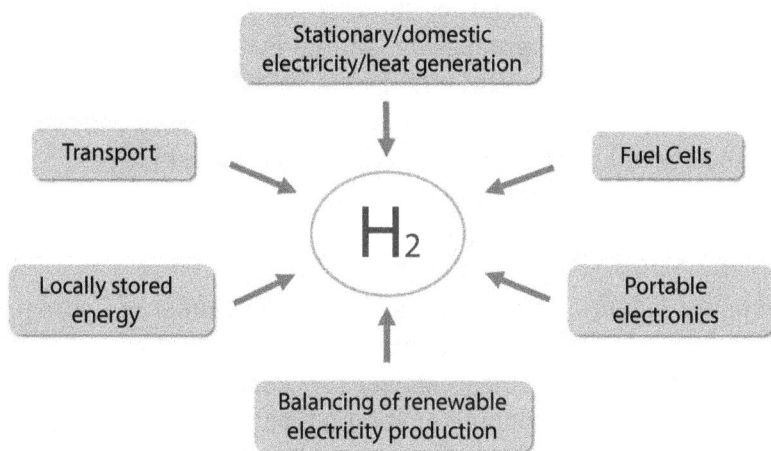

Fig. 3. Future Lines of Hydrogen Utilization

6.1 Refinery

The conventional market of hydrogen is mainly focused on refinery products. The principal processes that consume hydrogen in refineries are described below.

6.1.1 Hydrotreatment

Hydrotreatment is commonly used to remove oxygen, nitrogen, sulfur, and metals, but also to increase the hydrogen content of some heavy hydrocarbons and to saturate olefinic and aromatic ones, in the presence of a catalyst [22]. The increased demand for fuel has raised the interest of hydrotreatment, as it improves the cetane number, density, and smoke point.

Some hydrotreatment processes have specific functions. For example, hydrodenitrogenation (HDN) for removing nitrogen and hydrodesulfurization (HDS) for removing sulfur.

6.1.2 Hydrocracking

Certain residues from refinery processes, for instance, atmospheric and vacuum distillation columns, have low C/H, high viscosity, and heavy chains that need to be processed for their utilization. Hence, they demand hydrogenation to break the largest chains and improve the C/H ratio. Hydrocracking can resemble severe hydroprocessing. Unlike hydrotreatment, which is designed to eliminate pollutants from a feedstock, hydrocracking aims to break the bonds of heavy hydrocarbon chains and reduce the viscosity of hydrocarbons to obtain, e.g., diesel [22].

6.2 Ammonia and Urea Production

Ammonia production is the second higher demander of hydrogen in the current hydrogen economy [7]. Urea is a product of ammonia through the Bosch–Meiser process. The reactants are ammonia and CO_2, producing ammonium carbamate. By means of an ammonium carbamate decomposition reaction under slow endothermic conditions, urea is produced.

6.2.1 Ammonia Synthesis (Steam Reforming Method)

Ammonia has a great potential as a way to store energy, specifically talking about hydrogen storage [23]. The steam reforming method is the most widely used worldwide for the production of ammonia. Natural gas, mainly formed of 90% of methane, fuels the process.

This process consists of four different stages: desulfurization, reforming, purification, and synthesis of ammonia.

1. Desulphurization
   The first step is to remove the sulfur content of natural gas. Normally, gas distributors add sulfurs and organic compounds to odorize the mixture [22]. The main reactions are below.

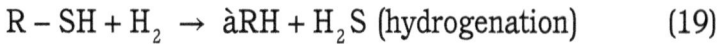

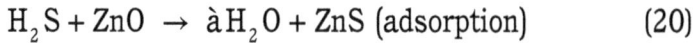

$$R - SH + H_2 \rightarrow \text{à} RH + H_2S \text{ (hydrogenation)} \tag{19}$$

$$H_2S + ZnO \rightarrow \text{à} H_2O + ZnS \text{ (adsorption)} \tag{20}$$

2. Reforming
   Once the natural gas is cleaned, it is subjected to a catalytic reforming with water vapor (cracking-ruptures of the CH_4 molecules). Natural gas is mixed with water vapor in the proportion (1:3)—(gas: water vapor) and is taken to the reforming process, which is carried out in two stages:
   Primary reformer
   The natural gas along with the steam passes through the equipment pipes where the following reactions take place:

$$CH_4 + H_2O \rightarrow CO + 3H_2 \, \Delta H = 206kJ/mol \tag{21}$$

$$CH_4 + 2H_2O \rightarrow CO_2 + 4H_2 \, \Delta H = 166kJ/mol \tag{22}$$

These reactions are strongly endothermic, reacting at 800°C and catalyzed by nickel oxide (NiO), thus, the formation of H_2 is favored. The heat needed in the primary reformer is supplied by burning a fuel with air.
Air is also introduced into the secondary reformer, but just to introduce nitrogen to the ammonia synthesis reactor.
Secondary reformer
The exhaust gas from the previous reformer is mixed with an air current in this second equipment, in this way the necessary N_2 for the stoichiometric synthesis gas, $N_2 + 3H_2$ is obtained. In addition, the combustion of methane takes place, reaching temperatures above 1,000°C.

$$CH_4 + 2O_2 \rightarrow CO_2 + 2H_2O \, \Delta H < < 0 \tag{23}$$

In summary, after these stages, the composition of the resulting gas is approximately N_2 (12.7%), H_2 (31.5%), CO (6.5%), CO_2 (8.5%), CH_4 (0.5%), Ar (0.1%). A 99% conversion of hydrocarbon has been achieved.

3. Purification
   The NH_3 production requires a high-purity synthesis gas. Therefore, CO and CO_2 cases must be removed first. For this purpose, three stages must be carried out: conversion of CO into CO_2, removal of CO_2 and methanation.
4. Ammonia synthesis
   The synthesis gas is compressed to a pressure of approximately 200 bars and taken to the reactor, where the production of ammonia takes place in a catalyst bed of Fe.

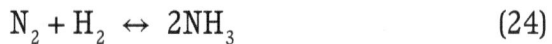

$$N_2 + H_2 \leftrightarrow 2NH_3 \tag{24}$$

The synthetic gas passes through the reactor for the first time, obtaining conversion efficiencies of between 14% and 15%. To improve efficiency, the gas is recirculated by first going through a process of separation of ammonia by condensation, and secondly, by a process of elimination of inert, since their accumulation is harmful to the process. Ammonia is stored in a cryogenic tank at −33°C [22].

In general, the synthesis of ammonia is a relatively clean process, but it consumes a high amount of energy.

6.3 Chemical Industry

The use of hydrogen in the chemical industry has a lower percentage of utilization than refineries and ammonia production, although its use is also considerably important. Figure 4

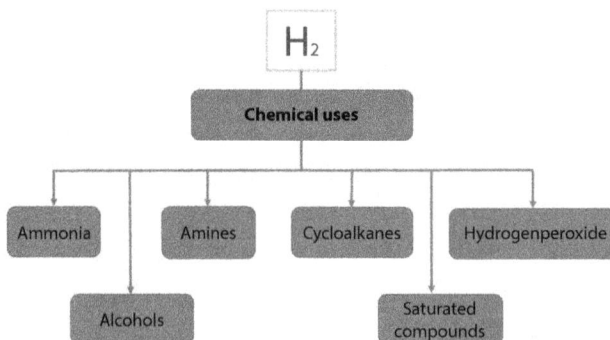

Fig. 4: Main Chemical Products That Demand Hydrogen in Their Production

illustrates the main uses of hydrogen for chemical compounds production.

6.4 *Fuel Cells*

The fuel cells are electrochemical systems in which the energy of the chemical reaction is directly converted in electricity. These kinds of systems work continuously if there is combustible and oxidant reacting inside the cell. Fuel cells consist of an anode and a cathode separated by an ionic conductor electrolyte. The hydrogen (sometimes ammonia or hydrazine) feed the anode, and the oxygen or air feed the cathode.

Its operating principle is based on the hydrogen oxidation, being the chemical reaction of this process, the reverse of the electrolysis. The chemical reaction is described below.

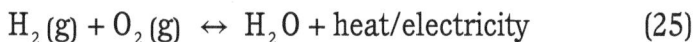

$$H_2\,(g) + O_2\,(g) \leftrightarrow H_2O + \text{heat/electricity} \qquad (25)$$

Fuel cells have an essential role in vehicle's future. The mobility sector has to overcome the environmental problems of conventional engines. The use of hydrogen as vehicle fuel clearly solve these problems, even if hydrogen is blue or gray [24]. However, the efficiency, the car range, and the operation and fuel cost still need to be optimized. The quick response and the operation temperatures, make PEM fuel cells (PEMFC) the most appropriate technology for vehicles [19].

Fuel cells can be also used for stationary heat/electricity generation. The stationary fuel cells have important advantages such as high efficiency, low carbon emissions, and higher flexibility. The last advantage is particularly important since stationary fuel cells can produce heat and electricity from different fuels obtaining an energy/heat ratio higher than the conventional cogeneration systems [25].

Moreover, despite fuel cells are already implemented and well studied, at small scale, the future of fuel cells is clearly promising. One of the most interesting research lines is the miniaturization of fuel cells. This technology will allow the substitution of fuel cells instead of small batteries for electronic devices such as laptops, mobile phones, radios, etc. However, this technology must overcome barriers such as low operation temperatures, fuel availability, and quick response and activation [26].

6.5 Space Uses

The combustion properties of hydrogen make it one of the favorites fuels for aerospace applications. A few of the most important characteristics are listed below [27].

- Higher release of energy
- Wide flammability limits (FLs)
- Short ignition time
- High diffusivity
- Thermal conductivity
- High heat capacity
- Low dynamic viscosity
- Higher specific impulse
- Clean combustion

However, there are important barriers that must be overcome to consolidate the use of hydrogen in the aerospace industry. For example, hydrogen storage capacity is limited to the necessity of increasing storage density, since hydrogen needs to be stored at high pressures. Moreover, the water vapor produced in the hydrogen-air reaction has a detrimental impact on the environment since the formation of condensation trails. Trails have a greater irradiative absorption strength, than CO_2, contributing to the greenhouse effect [28].

6.6 Enrichment of Combustible Mixtures with Hydrogen

Nowadays, the development of new combustion technologies capable of mitigating climate change effects seems to be essential. Several studies propose the enrichment of natural gas with hydrogen, in small concentrations, since it reduces greenhouse emissions like CO, CO_2, and NO_x. Moreover, the addition of hydrogen considerably improves flame stability.

6.6.1 *Natural Gas*

The enrichment of methane/natural gas with hydrogen has aroused great interest in the last decade for several reasons [29]. The main one is the improvement of the combustion characteristics in engines and heaters that directly imply a reduction in the greenhouse emissions.

Recent studies propose to use mixtures from 10% of hydrogen up to 30% to obtain the optimal combustion characteristics [30]. The concentration of hydrogen is a crucial parameter since the flame propagation, flame stability, the combustion speed, etc. depend on it [31].

Consequently, some authors, knowing the advantages mentioned above, propose the modification of the well-known "Power-To-Gas" avoiding the methanation process to inject hydrogen in the natural gas network [32].

6.6.2 *Biogas*

Biogas is produced in industrial plants by fermentation process. Lately, it has been demonstrated that the addition of hydrogen to such process can improve global efficiency by up to 95%. This method is under development, but it represents an interesting future research line [33]. Moreover, the CO_2 produced during the fermentation can be used to prevent explosions, as the hydrogen/methane mixture can become inert when diluted to a certain concentration with this gas [30].

7.0 Energy Storage

Hydrogen storage can be fundamentally divided into five big groups.

7.1 Storage of Pure Hydrogen

Nowadays, the storage of pure hydrogen in the gas or liquid phase is the only storage method used on a significant scale. Normally, hydrogen used in space is stored in the liquid phase, and underground storage is carried out with compressed gas. Both processes have high investments and process costs [34].

Storage of hydrogen in the gas phase always implies the compression of the gas to increase the density storage. The surface storage has pressures of about 100 bar (can be greater for refueling stations), while underground storage can reach pressures about 200 bar [35]. Despite the higher pressures of underground storage, surface storage is more expensive. Therefore, underground storage on a large scale is normally preferred [36].

The underground storage location is crucial depending on the stored gas, the storage time, the future applications, geological characteristics, or biological characteristics. The geological formation nature provides the first distinctions between reservoirs selection [37]. There are fundamentally two types:

- Poros media storage: Consists of filling the reservoir with hydrogen. The type of rock that can provide poros media storage must have permeability and high porosity, and, a feature most commonly found in rocks such as conglomerates, limestone, sandstones, etc. Typically, the reservoir types in sedimentary rocks are depleted hydrocarbon reservoirs and saline aquifers. Saline aquifers must have high salinity.
- Storage in engineered cavities: Consists of construct underground caverns with a well-defined geometry. The cavities may be created by dissolution of the rock. Typical rocks can be soft-rocks, such as shales or unfractured

limestone, hard-rocks, such as granites or basalts or salt. Cavities resulting from previous mining activities may also be used, depending on the mining technique and on the conditions of abandonment of the mine. Unlike poros media, engineered cavities are best suited to low porosity and low-permeability rocks to ensure that the stored fluid does not migrate to the surrounding formation. Therefore, unfractured or poorly fractured igneous and metamorphic rocks, or very low-permeability rocks are the target reservoirs. In recent years, research has focused on the possibility of using linings in the cavities to ensure containment of the fluids and allow for higher storage pressure [37].

Regarding geological factors, there are manly three types of sites; host rocks, salt caverns, and poros media. Depending just on geological criteria, the techno-economic feasibility and maturity of hydrogen storage can be summarized in Table 2.

Table 2: Review of Hydrogen Storage Maturity of Technology

Storage	Reservoirs	Maturity of Technology
Salt	Salt caverns	Proven technology, sparsely implemented
Host Rocks	Engineered cavities	Prospective technology, pre-commercial pilots conceptual designs
	Abandoned mines	Not applicable
Porous Media	Aquifers and traps	Prospective technology, precommercial pilots conceptual designs
	Depleted hydrocarbon reservoirs	Prospective technology, pre commercial pilots conceptual designs

This kind of storage is already being applied in some countries around the world, such as United States. Several geological locations have been suggested in recent years, but salt cavities are the most suitable. There are many reasons for using this kind of hydrogen storage, such as its low construction costs, fast injection and withdrawal rates, low leakage rates, and low risk of hydrogen contamination [38].

Despite hydrogen underground storage is a mature and consolidated technology, numerous projects around the word still pose uncertainties around this technology, such as changes in gas composition, volume losses, temperature increase, permeability reduction, acidification, or increase of corrosion in the surface installation [39]. Some of this projects and the anomalies detected are described below. These anomalies are caused by impure gas streams that can affect the geological bore integrity, reservoir properties, and simulate not desirable metabolisms. The impure gas streams can be the cushion gas present in the reservoir, or mixtures of gases injected in the reservoir. Some real examples are listed below [40].

- Hychico-BRGM pilot project: In the first stage, they produce hydrogen through wind power. Then, they started to inject the green hydrogen in a depleted gas reservoir. The deep of the reservoir is 815 m. During the test, the concentration of H2 reached the 10%. They observed that some hydrogen stored was converted to methane.
- Underground Sun Conversion (European consortium): This project demonstrates in a field test that the storage of natural gas with a mixture of 10% of hydrogen is technically feasible. They also observed changes in gas compositions in a sandstone reservoir in a depth of 1,000 m. After the first step of the project, they are now studying the chemical and biological processes that take place inside the reservoir.
- Ketzin (Germany): They observed an increase in H_2, CO_2, and CH_4 and a decrease of CO.
- Beynes (France): In this case, there are not any changes observed.

Departing on this results, a new research area has been created in order to understand the interaction between hydrogen and the reservoir characteristics. There are four main metabolisms playing an important role in hydrogen underground storage; methanogenesis, actogenesis, iron-reduction, and fermentation. The first one is the most important one regarding hydrogen interaction. The H_2 and the CO_2 are introduced into the reservoir via an injection well. During the injection, the metabolism of methanogenic archaea convert a portion of the injected mixture into methane and water. This process is called underground methanation.

Currently, researchers are focused on this phenomena, trying to define if the methanation reaction can be favored in underground conditions to generate methane by means of hydrogen storage.

The problems of underground hydrogen storage appear in those countries where there this geological location does not exist. One solution is to store hydrogen in underground metal vessels. The costs increase, but the storage stability also increases. Another solution is the pipeline storage. This method is currently used for the natural gas network, being more expensive in the construction of hydrogen pipelines.

Apart from compressed hydrogen, liquid hydrogen can be also stored. Hydrogen passes into liquid phase through liquefaction. Unlike natural gas, very high densities of stored hydrogen can be reached at atmospheric pressure (1 bar is 70 Kg/m³ [41]). Nevertheless, the low temperatures needed imply a large consumption of energy (−253°C at 1 bar). The liquefaction plant still poses enormous capital costs, being a significant part of the total cost of the hydrogen storage process. Moreover, the gas evaporation not only involves an energy loss but also implies safety problems such as the overpressure.

7.2 Adsorption of Hydrogen

Hydrogen absorption consists of benefiting the Van der Waals union between some materials and hydrogen. The materials must have a large specific surface area and the process works at low temperatures and elevated pressures due to the weakness of the Van der Walls' unions. Liquid nitrogen is the most used refrigerant by far, as its boiling point is −196°C [42]. In this case, the technology of hydrogen adsorption only exits on a laboratory scale.

At the moment, the best materials to adsorb hydrogen are certainly activated carbons and metal-organic frameworks (MOFs) that have an adsorption efficiency around 8%–10% [43]. The low efficiencies lead the research of this method to the use of additives to reach a hydrogen storage capacity higher than 40–50 Kg/m^3 at a temperature of −196°C [44].

However, one of the biggest barriers to the implementation of hydrogen adsorption is heat management. The hydrogen adsorption is exothermic. The heat generated must be removed to ensure the Van der Walls unions in the most effective way. Approximately, to remove the excess heat of just 1 kg of adsorbed hydrogen, it takes at least 10 kg of liquid nitrogen. This procedure increases operating costs and is essential to improve efficiency. Nevertheless, even with the temperature control, researches do not expect high efficiencies for this process in the coming years.

7.3 Metal Hydrides

Unlike hydrogen adsorption, the union between hydrogen and metal hydrides is much stronger. This kind of union allows the storage of hydrogen with higher density, even at ambient temperature and pressure. However, this fact implies the need for a higher quantity of energy to release the chemically bonded hydrogen.

There are two ways to obtain the hydrogen from metal hydrides, thermolysis and hydrolysis. The principle of thermolysis is the heating of the metal hydrides, and the reaction is endothermic and occurs in the solid phase, being necessary to work with elevated temperatures. Nevertheless, the hydrolysis uses water as a reagent, is exothermic and occurs in a solution under atmospheric conditions. Currently, the thermolysis method is the most studied in the literature, with only a few examples of hydrolysis. The most developed metal hydride is the sodium borohydride ($NaBH_4$) [45].

7.4   Chemical Hydrides

The principle of the hydrogen storage in chemical hydrides, or power-to-liquids storage, is similar to the storage in metal hydrides. However, there are a few points that make this method radically different. Chemical hydrides are fundamentally liquids that improve the viability of its transport and storage. Additionally, the most studied chemical hydrides for this kind of storage are methanol, formic acid, and ammonia, three widely produced chemicals. These two facts are advantageous since there is a current infrastructure available for their use in the medium term.

7.5   Hydrogen Injection on Natural Gas Network

Hydrogen can be stored through its transformation into methane using the methanation process. This application is called power-to-gas and has been extensively studied in recent years. However, the methanation process has low global efficiencies, so its utilization is currently limited. Interestingly, as described before, hydrogen can be mixed with methane or natural gas, as this mixture has better combustion properties. On this basis, some studies modify the well-known power-to-gas application, proposing direct injection of hydrogen in the natural gas network [32].

From the storage point of view, the main advantage of this method is the storage capacity of the current pipeline network of natural gas. This capacity can be considered almost unlimited, as the natural gas amount is huge. Nevertheless, the hydrogen content is commonly limited to guarantee the fuel quality and to avoid problems in energy production processes.

8.0 Safety Use of Hydrogen and Hydrogen Mixtures

Gas explosions have caused serious and lethal accidents in the industry since the beginning of the industrial era. Nowadays, gas explosions still occur and experience demonstrates that the prevention of leakages, the formation of explosive clouds, and the ignition sources are not enough to reduce the explosion risk. Consequently, a deep knowledge of the properties of flammable gases is essential in order to manage the risk of explosion and to implement safety measures [46].

Hydrogen is one of the most easily flammable gases and its physical properties as a result of its small molecule, such as lightness and diffusivity, make it quite dangerous, since leakages are common when handled. Indeed, the highly flammable nature of hydrogen still poses major safety problems today, due to the formation of explosive atmospheres. The capability to alleviate these safety problems departs from the hydrogen flammability characterization.

To characterize gas flammability, different groups of parameters can be determined:

(a) Ignition characteristics:
Minimum Ignition Temperature (MIT): Minimum temperature to ignite the mixture, depending on the concentration.
Minimum Ignition Energy (MIE): Depending on the concentration, MIE is the lowest energy to initiate the explosion. It is measured by means of a capacitive discharge.
Minimum Ignition Current (MIC): This parameter is related to MIE, being the MIC easier to evaluate. MIC is the minimum current to produce the ignition in a standardized equipment with an inductive circuit. Values are expressed by comparison with the MIC of methane.

(b) Combustion characteristics:
Lower FL (LFL): LFL is the lower limit of the explosive range of a fuel-oxidizing agent mixture, expressed as

the volume percentage in its mixture with air. Upper FL (UFL): UFL is the upper limit of the explosive range of a fuel-oxidizing agent mixture, expressed as the volume percentage in its mixture with air.

Flash Point (FP): FP is the temperature at which the equilibrium of the liquid–vapor phase equilibrium reaches the LFL. This property is only suitable for flammable substances that are liquid under atmospheric conditions.

(c) Explosion behavior:

Maximum Explosion Pressure (Pmax): it is the highest explosive pressure in a closed vessel when an explosion occurs.

Maximum rate of pressure rise (dp/dt max): is the relation of the pressure increase to the time interval needed to produce such increment due to an explosion.

Maximum Experimental Safe Gap (MESG): It is the maximum safety gap able to prevent the explosion propagation in standardized equipment with a sealing length of 12.5 m Table 3).

Table 3: Flammability Parameters of Hydrogen [46]

LFL (%V)	UFL (%V)	MIE (mJ)	MIC	MIT (°C)	FP (°C)	MESG (mm)	Subgroup
4.0	75.6	20	0.27	560	–	0.29	C

All flammable gases are classified into two groups: Group I is for mine gas, that is firedamp. Group II is for all the industrial gases.

Additionally, Group II of flammable gases can be classified depending on the MIE, MIC, and MESG since there is a certain correlation between them. Table 4 shows the subgroup classification, where flammability increases from IIA to IIC, being IIC the most easily ignitable gases.

Table 4: Subgroup Classification of Flammable Gases [46]

Subgroup	MIE (µm)	MIC/MIC(CH4)	MESEG (mm)
A	205 < e	0.8 < 1	0.9 < i
B	96 < e < 250	0.45 < 1 < 0.8	0 < e < 0.9
C	e < 96	1 < 0.45	i < 0.5

Flammable gases can be also classified according to their MIT values, so that equipment intended to be used in the presence of a flammable gas must have its surface temperature according to its MIT value. Thus, such equipment will be marked by the manufacturer indicating its Temperature Class. Table 5 shows the classification.

Table 5: Classification of Flammable Gases Depending on the Surface Temperature

Temperature Class	T (°C)
T6	85
T5	100
T4	135
T3	200
T2	300
T1	450

However, hydrogen is not always used as a pure gas, since it participates in several industrial processes and comes into contact with numerous compounds. This fact implies that the explosive atmosphere is not always formed by hydrogen-air, being possible to find it mixed with pure oxygen, chlorine, NO and NO_2, and many other gases. When a flammable gas is mixed with different gases, some properties change, such as the LFL and uUFL, as shown in Table 6.

Table 6: FLs of Hydrogen When Mixed With Other Gases

	Hydrogen FL (%)
Air	4.0–75.6
Oxygen	3.9–95.8
Chlorine	3.5–89.0

Nevertheless, if hydrogen is mixed with inert gases such as nitrogen or carbon dioxide, all the flammability properties decrease and the risk of explosion can even be completely avoided. Indeed, inert gases are commonly used in industry to reduce the explosion range of flammable mixtures and to control safe concentrations in the working environment. When an inert gas is added to an explosive mixture, such as air-hydrogen, the limiting oxygen concertation (LOC) is the threshold value at which the atmosphere results completely inerted. By controlling this parameter, it is possible to avoid the formation of explosive atmospheres. The minimum volume of inert gas needed to make the atmosphere inert is shown in Table 7. The most used inert gas in industry is the nitrogen. However, the carbon dioxide has a great inert power when mixed with hydrogen-air mixtures.

Table 7: LOC of Hydrogen When Is Diluted With Different Inert Gases

LOC (%V)	
N_2	CO_2
71	57

Sometimes, the risk of explosion cannot be completely removed. Therefore, it is necessary to apply prevention measures intended to reduce the probability of activation of ignition sources and also to develop protection measures to mitigate the effects of an explosion.

9.0 Green Hydrogen Storage and Utilization

Once the state of the art has been reviewed, the authors of this chapter propose two different solutions for hydrogen storage and utilization. As already mentioned, the integration of renewable energy sources into the current electricity grid, unbalances the demand-generation curves. The intermittence of renewable resources such as solar and wind energy complicates the alienation between both curves.

The balance of the demand-generation curves can be reached by means of energy storage. The use of electric batteries is limited by the storage capacity of current technology and raw materials reservoirs. Nevertheless, energy can be stored through other energy vectors such as hydrogen.

9.1 Solution A

Pure hydrogen can be stored above or underground as described in Section 7.1, although one of the most promising ways to store this gas is by injecting it in the natural gas network. Therefore, following the "hydrogen square," the proposed solution involves the main key pillars of hydrogen: production, utilization, storage, and safety, as described in Fig. 5.

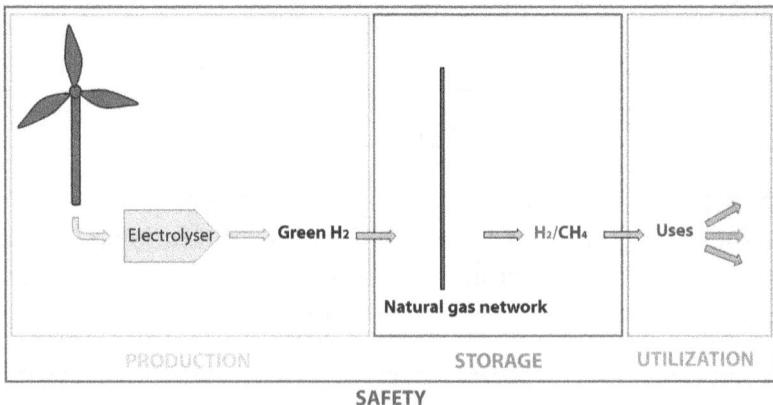

Fig. 5: Scheme of the Green Energy Storage System

The excess energy generated by, for instance, wind energy can be used directly to feed electrolysers. The electrolysers transform water into hydrogen. Depending on the electrolyser used in the process, the hydrogen stream might or might not be required to undergo a purification process. When hydrogen reaches a purity of at least 99%, it can be injected in the natural gas network.

This proposed solution allows the generation of green energy, which can be stored in a high capacity network. Moreover, the infrastructures are already in place. The mixture between hydrogen and natural gas has great advantages when used in combustion processes as described in Section 6.6.

However, the hydrogen injection technique needs to accomplish the quality standards of natural gas distribution and commercialization. The percentage of hydrogen injection depends on the gas volume of the network connection point of transport/distribution, intended to use for the injection. Therefore, the calculation of the correct percentage will rely on the gas system of each country and the location of the selected connection point. Moreover, it is crucial to guaranty an enriched gas with equal calorific value than the natural gas, since an excess on the calorific value can unbalance the combined cycle plants. The reason is that hydrogen has greater calorific value than natural gas, which is reflected in a heat increment. If the temperature increase beyond the work parameters of the plant, the use of this fuel can affect seriously the materials, and the efficiency of the cycle. Applying simple calculations as described in Section 9.2, it is possible to see how hydrogen enrichment affects the calorific value. In fact, the calculations show the addition of 10% of hydrogen enhances the lower calorific value by almost 15%. In conclusion, the hydrogen injection in the natural gas network must take into account all these parameters, particularly the requirements of national regulations.

Below are listed the main advantages of this storage system, depending on the four pillars of hydrogen implementation:

1. *Production:* This system produces green hydrogen. System should be based on intermittent and unpredictable renewable energy such as wind or photovoltaic systems.
2. *Storage:* The capacity of storage of the natural gas network is almost unlimited due to the extensive network of existing pipelines. Moreover, this system allows for the storage of excesses of green energy and balancing the demand-production curves.
3. *Utilization:* Natural gas has numerous final uses in the current energy system. Consequently, by mixing hydrogen with natural gas, the uses of hydrogen are much more than those of pure hydrogen–at least until current the technology reaches the needed maturity.
4. *Safety:* Natural gas and hydrogen are flammable gases that can explode when mixed with air. The flammability of hydrogen is greater than the flammability of natural gas, being this one of the main problems of hydrogen implementation. However, the mixture of hydrogen with natural gas considerably reduces the risk of explosion compared with pure oxygen.

Fig. 5 describes the process of green energy storage proposed in this chapter.

9.2   Solution B

However, chemical energy storage can be carried out through small distributed storage, which contributes to the use of surplus energy in off-peak hours. As happens with electric car charges in off-peak hours, small-scale production of hydrogen can contribute to the use of surplus energy. In this section, a solution of hydrogen utilization in small scale is provided.

The energy consumption is growing globally day by day since the beginning of the industrial era. The search for new and clean energy sources quickly reached the fuel sector. This sector is currently driven by oil products. However, the lack of resources and the necessity to fight against climate change have boosted great interest in renewable fuels like biofuels or biogas.

Biofuels are a controversial issue as it competes with land use. By contrast, biogas can be produced by urban, farming, and livestock waste, making it a potential renewable fuel for the medium-term future.

Biogas is an alternative fuel originated from the organic matter degradation under anaerobic conditions. Depending on the raw material, the composition of biogas can vary considerably. Methane and carbon dioxide are the main gases present in this renewable gas. Its high CO_2 content is precisely the main reason of the low calorific value (LCV) of biogas.

The calorific value is a critical parameter of fuels. This parameter reflects the detached energy from the combustion per unit of mass. This energy released in the form of heat is related to combustion efficiencies in engines, boilers, etc. Hence, the LCV of biogas prevents its use when biogas has high CO_2 content. Therefore, currently biogas is purified before using it, consequently the price of biogas increases due to the purification process and making it less competitive [47].

As introduced in Section 6.6., the enrichment of natural gas with hydrogen enhances its combustion characteristics. On this basis, well consolidated in literature [48–51], the authors of this chapter propose to enrich biogas with hydrogen to enhance certain biogas properties. Biogas enriched with hydrogen may have similar combustion characteristics than natural gas, even avoiding the purification process of carbon dioxide.

This mixture can be used for natural gas applications such as self-consumption boilers. If hydrogen is produced through renewable energies during off-peak hours, it is possible to store energy and to produce a renewable fuel at the same time (Fig. 6).

The study of the effect of hydrogen addition to biogas, on the combustion process, is complex due to the dual interaction of CO_2 and H_2. Hydrogen has greater calorific value than methane, butane, propane, etc., hence, the addition of hydrogen to biogas will raise the calorific value of the mixture. Nevertheless, CO_2 reduces the temperature during the fuel combustion that can be an opposite effect on hydrogen addition. In order to deeply understand the combustion behavior of the mixture is essential the development of experimental tests. But, as a first approach,

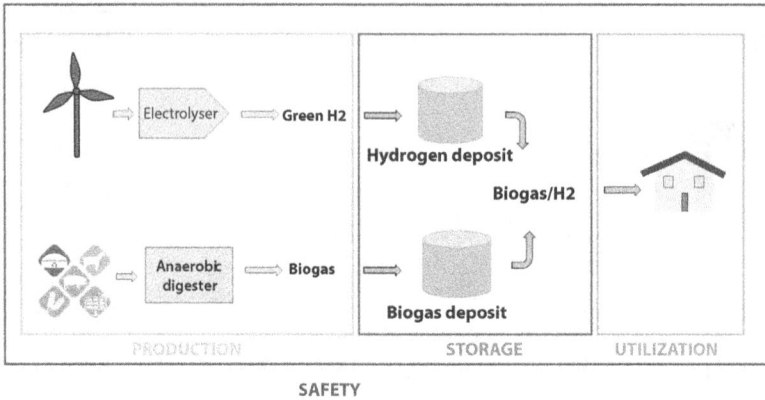

Fig. 6: Scheme of the Green Energy Storage System

in order to evaluate the effect of hydrogen addition the authors perform a theoretical study of the calorific value of different mixtures.

According to standard EN ISO 6976 [52], the high calorific value (HCV) and the low calorific value (LCV) of a gas mixture must be calculated by applying the following formula:

$$(Hc)_G = (Hc)_G^0 = \sum_{j=1}^{N} x_j \left[(Hc)_G^0 \right]_j \qquad (26)$$

Where: $(Hc)_G$ is the molar value of the calorific value of the ideal gas mixture; $(Hc)_G^0$ is the molar value of the calorific value of the real gas mixture; $\left[(Hc)_G^0 \right]_j$ is the ideal molar value of the calorific value of the j component; x_j is the molar fraction of the j component.

The LCV data vary in the literature depending on the source. Table 8 resumes the LCVs for the gases used to develop this theoretical study.

Natural gas and biogas are gases complex mixture. In Table 9, a list of the gases that make up natural gas can be seen. The composition of each component of the fuel ranges depending on the origin [54,55]. Nevertheless, the commercialization of these fuels must accomplish the requirements of the transport and distribution

Table 8: LCVs for Fuel Gases [53].

Gas	LCV (kJ/kg)
Methane	50.000
Ethane	47.510
Propane	46.200
Butane	45.790
Hydrogen	120.011
Natural gas	39.900
Carbon monoxide	10.160

standards of each country. These standards regulate the maximum content of some of these gases to guarantee the quality of the fuel.

Table 9: Gases Composition for Natural Gas and Biogas [55,56]

Natural Gas	Formula	% Molar	Biogas	Formula	% Molar
Methane	CH_4	91.46	Methane	CH_4	50–70
Ethane	C_2H_6	3.58	Carbon dioxide	CO_2	30–50
Propane	C_3H_8	1.45	Carbon monoxide	CO	0.10
Butane	C_4H_{10}	0.65	Oxygen	O_2	0.10
Carbon dioxide	CO_2	1.76	Nitrogen	N_2	0.50
Nitrogen	N_2	0.80	Hydrogen sulfate	H_2S	0.10
Others	–	0.3	Others	–	0.80

The biogas composition is a crucial parameter in order to define the calorific value and the combustion parameters. The methane and carbon dioxide content can vary by 20% due to the organic matter used for biogas production. Depending on the methane content, biogas can be rich, medium, or poor. This study proposes three different biogases compositions as exposed in Table 10.

Table 10: Composition of Poor, Medium, and Rich Biogas Defined in This Study

Gas	Formula	Poor % Molar	Medium % Molar	Rich % Molar
Methane	CH_4	49.60	59.60	69.60
Carbon dioxide	CO_2	49.60	39.60	29.60
Carbon monoxide	CO	0.10	0.10	0.10
Oxygen	O_2	0.10	0.10	0.10
Nitrogen	N_2	0.50	0.50	0.50
Hydrogen sulfate	H_2S	0.10	0.10	0.10

The calculation of LCV is carried out considering a not purified biogas. Figure 7 compares the LCV of standard natural gas, biogas ($CH_4:CO_2$/ 59:20), and biogas ($CH_4:CO_2$/59:20) enriched with 20% of hydrogen. The results reflect a difference of calorific value

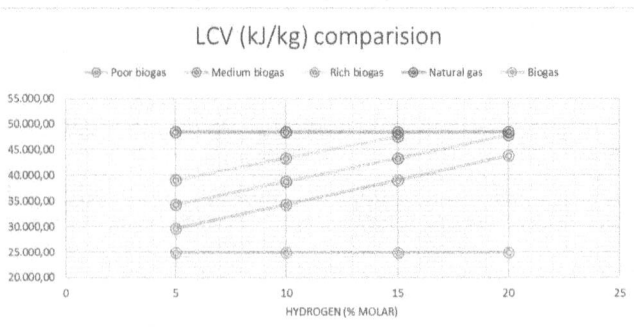

Fig. 7: LCV Comparison of Standard Natural Gas, Biogas, and Biogas Enriched With Hydrogen

between natural gas and biogas of 28.42%, while biogas enriched with hydrogen enhance its calorific value around 10%.

Then, to understand how affect the enrichment of biogas with hydrogen to the caloric value of biogas, we have calculated the calorific value of the three biogas compositions of Table 10, caring the percentage of hydrogen enrichment in 5%, 10%, 15%, 20%. Figure 8 plot the results.

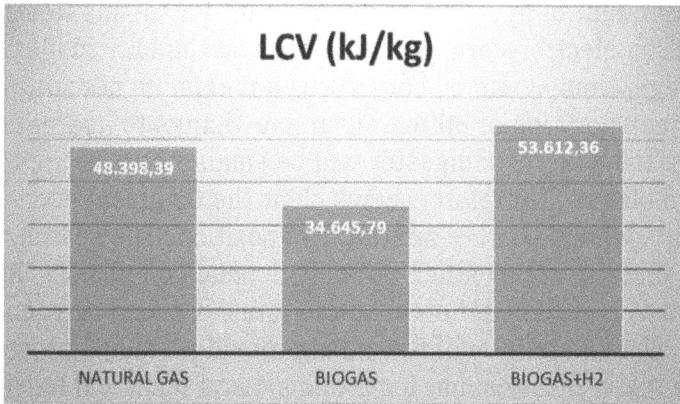

Fig. 8: LCV Comparison of Poor, Medium, and Rich Biogas When Hydrogen is Added in Different Percentages

Natural gas and biogas in Fig. 8 are constant as they are not enriched with hydrogen. The LCV of poor, medium, and rich biogas increases almost linearly with hydrogen addition. This effect has been observed in experimental investigations regarding different combustion parameters such as the laminar burning velocity or the improvement of radical H^+ in the reactions taking part during the combustion process [57].

The improved fuel kinetics and enhancement of adiabatic flame temperature can be related to hydrogen addition, as it is liked to LCV. Nevertheless, carbon dioxide can contribute to a decrease in flame temperature. Similarly, the hydrogen enrichment decreases the flame stability of biogas fuel effectively, while the increased CO_2 content improves flame stability of biogas–hydrogen fuel [57]. All these facts reflect the necessity to research more deeply the combined effects of CO_2 and H_2 on the combustion parameters.

Biogas composition is again a crucial parameter for LCV comparison. Depending on the type of biogas, the quantity of hydrogen needed to reach the LCV of natural gas increases. Rich biogas reaches the LCV of gas natural just by adding a 15% of hydrogen. However, medium biogas needs at least 20% of hydrogen, while poor biogas needs more than 25%.

The amount of hydrogen is determinant for the installation design. The electrolysers sizing depends on the enrichment percentage, thus, it can be considered a maximum percentage of hydrogen addition of 20%. Conversely, thanks to hydrogen addition, the anaerobic digester will be smaller comparing to standard biogas. Therefore, seems clear the technical/economic feasibility depends on the mixture composition. The search for the optimal mixture composition not only depends on physicochemical parameters and combustion characteristics but also on economic factors. The elimination of the carbon dioxide removal process must be taken into account for economic and technical feasibility.

Under the view of the results of Fig. 8, the advantages of hydrogen addition seem clear talking about calorific value. However, despite its high importance regarding biogas distribution and commercialization, this parameter is not determinant in order to confirm the enhance of enriched biogas properties.

10.0 Conclusions

Nowadays hydrogen is a gas commonly used in diverse chemical production, storage, distribution, transportation, and end-use applications. The main hydrogen-consuming sectors are the chemical industry, the space sector, power generation, and energy distribution. Nevertheless, in addition to the current uses of hydrogen, this gas seems to be one of the energy vectors of the future, since its versatility and great potential make it a viable solution to current environmental problems.

The participation of renewable energies in the electric grid presents stability problems that can be solved through the use of hydrogen. The intermittent production of solar and wind energy disrupts the stability of the current electric system, being necessary the energy storage. The answer to this problem lies in the storage of excess power in off-peak hours. Such excess power can be used to dissociate water through electrolysis, producing hydrogen as an energy vector. The produced hydrogen can be used in numerous ways, as described in this chapter, being possible the implementation of a hydrogen economy.

However, its use also involves disadvantages that must be taken into account. Hydrogen leakage poses a risk to the ozone layer, being quite difficult to control because of its lightness. Moreover, the hydrogen flammability implies working under controlled conditions and taking safety measures to mitigate the risk of explosions. Even when controlling all the parameters there are always some risks associated with its use.

The creation of a hydrogen network can provide the next generation with a sustainable future to overcome the current problems, but it is still needed to solve and improve the main disadvantages proposed in this chapter. Nevertheless, this chapter proposes a solution that can be implemented in current infrastructures by means of a mixture of hydrogen and natural gas, and a mixture of hydrogen and biogas. Hydrogen injection into the natural gas network has considerable advantages and allows the storage of green hydrogen as an energy vector. Moreover, small distributed storage of hydrogen can help to grid balance and to generate new opportunities in rural areas as the enrichment of biogas with hydrogen. The global objectives of emission reduction inevitably oblige countries to boost new clean fuels, therefore, the valorization of biogas as a fuel that can be consumed directly seems essential nowadays. Furthermore, this mixture allows for the efficient use of agricultural and livestock wastes.

Acknowledgments

The authors would like to thank Sofía Nawrot Flores for her contribution to the work as a translator, editor and proof-reader.

References

(1) Dincer I. Green methods for hydrogen production. *Int J Hydrog Energy*. 2012;37(2):1954–1971. doi:10.1016/j. ijhydene.2011.03.173

(2) Nikolaidis P, Poullikkas A. A comparative overview of hydrogen production processes. *Renew Sustain Energy Rev*. 2017;67:597–611. doi:10.1016/j.rser.2016.09.044

(3) Acar C, Dincer I. The potential role of hydrogen as a sustainable transportation fuel to combat global warming. *Int J Hydrog Energy*. 2020;45(5):3396–3406. doi:10.1016/j. ijhydene.2018.10.149

(4) Yang X, Han D, Zhao Y, Li R, Wu Y. Environmental evaluation of a distributed-centralized biomass pyrolysis system: a case study in Shandong, China. *Sci Total Environ*. 2020;716:136915. doi:10.1016/j.scitotenv.2020.136915

(5) Dawood F, Anda M, Shafiullah GM. Hydrogen production for energy: an overview. *Int J Hydrog Energy*. 2020;45(7):3847-3869. doi:10.1016/j.ijhydene.2019.12.059

(6) Widera B. Renewable hydrogen implementations for combined energy storage, transportation and stationary applications. *Therm Sci Eng Prog*. 2020;16:100460. doi:10.1016/j. tsep.2019.100460

(7) Chaubey R, Sahu S, James OO, Maity S. A review on development of industrial processes and emerging techniques for production of hydrogen from renewable and sustainable sources. *Renew Sustain Energy Rev*. 2013;23:443-462. doi:10.1016/j.rser.2013.02.019

(8) Steinberg M, Cheng HC. Modern and prospective technologies for hydrogen production from fossil fuels. *Int J Hydrog Energy*. 1989;14(11):797-820. doi:10.1016/0360-3199(89)90018-9

(9) Lam MK, Loy ACM, Yusup S, Lee KT. Biohydrogen production from algae. In: *Biohydrogen*. Elsevier: 2019;37(2012):219-245. doi:10.1016/b978-0-444-64203-5.00009-5

(10) Sharma A, Arya SK. Hydrogen from algal biomass: a review of production process. *Biotechnol Rep*. 2017;15:63-69. doi:10.1016/j.btre.2017.06.001

(11) Barbosa MJ, Rocha JMS, Tramper J, Wijffels RH. Acetate as a carbon source for hydrogen production by photosynthetic

bacteria. *J Biotechnol.* 2001;85(1):25-33. doi:10.1016/
S0168-1656(00)00368-0

(12) Eroglu I, Aslan K, Gündüz U, Yücel M, Türker I. Substrate
consuption rates for hydrogen production by Rhodobacter
sphaeroides. *J Biotechnol.* 1999;27:1315-1329.

(13) Kumar G, Shobana S, Nagarajan D, et al. Biomass based
hydrogen production by dark fermentation — recent trends
and opportunities for greener processes. *Curr Opin Biotechnol.*
2018;50:136-145. doi:10.1016/j.copbio.2017.12.024

(14) Tao Y, Chen Y, Wu Y, He Y, Zhou Z. High hydrogen yield from a
two-step process of dark- and photo-fermentation of sucrose.
Int J Hydrog Energy. 2007;32(2):200-206. doi:10.1016/j.
ijhydene.2006.06.034

(15) Chi J, Yu H. Water electrolysis based on renewable energy
for hydrogen production. *Cuihua Xuebao/Chin J Catal.*
2018;39(3):390-394. doi:10.1016/S1872-2067(17)62949-8

(16) Shiva Kumar S, Himabindu V. Hydrogen production by PEM
water electrolysis – a review. *Mater Sci Energy Technol.*
2019;2(3):442-454. doi:10.1016/j.mset.2019.03.002

(17) Buttler A, Spliethoff H. Current status of water electrolysis for
energy storage, grid balancing and sector coupling via power-
to-gas and power-to-liquids: a review. *Renew Sustain Energy Rev.*
2018;82:2440-2454. doi:10.1016/j.rser.2017.09.003

(18) Zhang X, Song Y, Wang G, Bao X. Co-electrolysis of CO_2 and
H_2O in high-temperature solid oxide electrolysis cells: recent
advance in cathodes. *J Energy Chem.* 2017;26(5):839-853.
doi:10.1016/j.jechem.2017.07.003

(19) Paul B, Andrews J. PEM unitised reversible/regenerative
hydrogen fuel cell systems: state of the art and technical
challenges. *Renew Sustain Energy Rev.* 2017;79:585-599.
doi:10.1016/j.rser.2017.05.112

(20) García-Camprubí M, Izquierdo S, Fueyo N. Challenges
in the electrochemical modelling of solid oxide fuel and
electrolyser cells. *Renew Sustain Energy Rev.* 2014;33:701-718.
doi:10.1016/j.rser.2014.02.034

(21) Saba SM, Müller M, Robinius M, Stolten D. The investment
costs of electrolysis – A comparison of cost studies from the
past 30 years. *Int J Hydrog Energy.* 2018;43(3):1209-1223.
doi:10.1016/j.ijhydene.2017.11.115

(22)   Abdin Z, Zafaranloo A, Rafiee A, Mérida W, Lipiński W, Khalilpour KR. Hydrogen as an energy vector. *Renew Sustain Energy Rev*. 2020;120:109620. doi:10.1016/j.rser.2019.109620

(23)   Mah AXY, Ho WS, Bong CPC, et al. Review of hydrogen economy in Malaysia and its way forward. *Int J Hydrog Energy*. 2019;44(12):5661-5675. doi:10.1016/j.ijhydene.2019.01.077

(24)   Wang Y, Ruiz Diaz DF, Chen KS, Wang Z, Adroher XC. Materials, technological status, and fundamentals of PEM fuel cells – a review. *Mater Today*. 2020;32:178-203. doi:10.1016/j. mattod.2019.06.005

(25)   Ijaodola OS, El- Hassan Z, Ogungbemi E, et al. Energy efficiency improvements by investigating the water flooding management on proton exchange membrane fuel cell (PEMFC). *Energy*. 2019;179:246-267. doi:10.1016/j.energy.2019.04.074

(26)   Tanneru HK, Kuruvinashetti K, Pillay P, Rengaswamy R, Packirisamy M. Feasibility studies of micro photosynthetic power cells as a competitor of photovoltaic cells for low and ultra-low power iot applications. *Energ*. 2019;12(9):5-9. doi:10.3390/en12091595

(27)   Cecere D, Giacomazzi E, Ingenito A. A review on hydrogen industrial aerospace applications. *Int J Hydrog Energy*. 2014;39(20):10731-10747. doi:10.1016/j.ijhydene.2014.04.126

(28)   Haglind H, Hasselrot A, Singh R. Potential of reducing the environmental impact of aviation by using hydrogen Part I: background, prospects and challenges. *Aeronaut J*. 2006;2990:533-540.

(29)   Amez I, Castells B, Garcia-Torrent J, Medic L. Influence of the addition of hydrogen on the flammability intervals of the methane, air and CO_2 mixtures. Proceedings of the Air and Waste Management Association's Annual Conference and Exhibition, AWMA; June 2019; Varna, Bulgaria.

(30)   Molnarne M, Schroeder V. Hazardous properties of hydrogen and hydrogen containing fuel gases. *Process Saf Environ Prot*. 2019;130:1-5. doi:10.1016/j.psep.2019.07.012

(31)   Li T, Hampp F, Lindstedt RP. Experimental study of turbulent explosions in hydrogen enriched syngas related fuels. *Process Saf Environ Prot*. 2018;116:663-676. doi:10.1016/j. psep.2018.03.032

(32)   de Santoli L, Paiolo R, Lo Basso G. Energy-environmental experimental campaign on a commercial CHP fueled with H2NG

blends and oxygen enriched air hailing from on-site electrolysis. *Energy*. 2020;195:116820. doi:10.1016/j.energy.2019.116820

(33) Gómez Camacho CE, Ruggeri B, Mangialardi L, Persico M, Luongo Malavé AC. Continuous two-step anaerobic digestion (TSAD) of organic market waste: rationalising process parameters. *Int J Energy Environ Eng*. 2019;10(4):413-427. doi:10.1007/s40095-019-0312-1

(34) Tarkowski R. Underground hydrogen storage: characteristics and prospects. *Renew Sustain Energy Rev*. January 2019;105:86–94. doi:10.1016/j.rser.2019.01.051

(35) Wolf E. Large-scale hydrogen energy storage.pdf. In: Moseley PT, Garche J, eds. *Electrochemical Energy Storage for Renewables Sources and Grid Balancing*. Elsevier; 2015:129–142; ISBN 9780444626165, https://doi.org/10.1016/B978-0-444-62616-5.00009-7.

(36) Crotogino F, DonadeiS, Bünger U, Landinger H. Larger scale hydrogen storage securing future energy supplies. In: Stolten D, Grube T, eds. *18th World Hydrogen Energy Conference 2010 - WHEC 2010 Parallel Sessions Book 4: Storage Systems / Policy Perspectives, Initiatives and Cooperations Proceedings of the WHEC*, May 16-21. 2010, *Essen Schriften des Forschungszentrums J̧lich / Energy & Environment, Vol. 78-4 Institute of Energy Research - Fuel Cells (IEF-3) Forschungszentrum J̧lich GmbH*. 2010:411–429; Zentralbibliothek, Verlag; ISBN: 978-3-89336-654-5.

(37) Matos CR, Carneiro JF, Silva PP. Overview of large-scale underground energy storage technologies for integration of renewable energies and criteria for reservoir identification. *J Energy Storage*. 2019;21:241-258. doi:10.1016/j.est.2018.11.023

(38) Kruck O, Crotogino F, Prelicz R, Rudolph T. Overview on all known underground storage technologies for hydrogen. 2013. http://www.sciencedirect.com/science/article/pii/S0920410514003374

(39) Panfilov M . Underground storage of hydrogen: in situ self-organisation and methane generation. *Transp Porous Media*. 2010;85:841-865. 10.1007/s11242-010-9595-7. http://www.sciencedirect.com/science/article/pii/S1364032120300411

(40) Strobel G, Hagemann B, Huppertz TM, Ganzer L. Underground bio-methanation: concept and potential. *Renew Sust Energ Rev*. 2020;123:109747. ISSN 1364-0321, https://doi.org/10.1016/j.

rser.2020.109747; http://www.sciencedirect.com/science/article/pii/S1364032120300411

(41) Godula-Jopek A, Jehle W, Wellnitz J. Storage of pure hydrogen in different states. In: *Hydrogen Storage Technologies*. 2012:97-170. doi:10.1002/9783527649921.ch4

(42) Berenguer-Murcia Á, Marco-Lozar JP, Cazorla-Amorós D. Hydrogen storage in porous materials: status, milestones, and challenges. *Chem Rec*. 2018;18(7):900-912. doi:10.1002/tcr.201700067

(43) Blankenship TS, Balahmar N, Mokaya R. Oxygen-rich microporous carbons with exceptional hydrogen storage capacity. *Nat Commun*. 2017;8(1):1545. doi:10.1038/s41467-017-01633-x

(44) García-Holley P, Schweitzer B, Islamoglu T, et al. Benchmark study of hydrogen storage in metal-organic frameworks under temperature and pressure swing conditions. *ACS Energy Lett*. 2018;3(3):748-754. doi:10.1021/acsenergylett.8b00154

(45) Demirci UB. About the technological readiness of the H_2 generation by hydrolysis of B(−N)−H compounds. *Energy Technol*. 2018;6(3):470-486. doi:10.1002/ente.201700486

(46) García Torrent J. *Seguridad Industrial en Atmósferas Explosivas*. Madrid, Spain: Laboratorio Oficial J.M. Madariaga; 2003.

(47) Putmai N, Jarunglumlert T, Prommuak C, et al. Economic analysis of swine farm management for the enhancement of biogas production and energy efficiency. *Waste Biomass Valor*. 2020;11:5635-5645. doi:10.1007/s12649-020-00989-4

(48) Cellek MS, Pınarbaşı A. Investigations on performance and emission characteristics of an industrial low swirl burner while burning natural gas, methane, hydrogen-enriched natural gas and hydrogen as fuels. *Int J Hydrog Energy*. 2018;43(2):1194-1207. doi:10.1016/j.ijhydene.2017.05.10

(49) Rajpara P, Shah R, Banerjee J. Effect of hydrogen addition on combustion and emission characteristics of methane fuelled upward swirl can combustor. *Int J Hydrog Energy*. 2018;43(36):17505-17519. doi:10.1016/j.ijhydene.2018.07.111

(50) Wang D, Ji C, Wang S, Yang J, Tang C. Experimental investigation on near wall ignited lean methane/hydrogen/air flame. *Energy*. 2019;168:1094-1103. doi:10.1016/j.energy.2018.11.115

(51) Wang Q, Zhao Y, Wu F, Bai J. Study on the combustion characteristics and ignition limits of the methane homogeneous

charge compression ignition with hydrogen addition in micro-power devices. *Fuel*. 2019;236:354-364. doi:10.1016/j.fuel.2018.09.010

(52) EN ISO 6976:2016. Natural gas-Calculation of calorific values, density, relative density and Wobbe index from composition. 2017. www.iso.org.

(53) Solares C, ed. Gas turbines: a handbook of air, land and sea applications. 2nd ed. Oxford, UK:Butterworth-Heinemann; 2008. https://doi.org/10.1016/B978-075067969-5.50001-6, http://www.sciencedirect.com/science/article/pii/B9780750679695500016

(54) Hosseini SE, Bagheri G, Khaleghi M, Abdul Wahid M. Combustion of biogas released from palm oil mill effluent and the effects of hydrogen enrichment on the characteristics of the biogas flame. *J Combustion*. 2015;2015:612341. doi:10.1155/2015/612341

(55) Ryckebosch E, Drouillon M, Vervaeren H. Techniques for transformation of biogas to biomethane. *Biomass Bioenergy*. 2011;35(5):1633–1645. doi:10.1016/j.biombioe.2011.02.033

(56) Mezni M. *LPG Recovery Unit Optimization*. New Mexico Institute of Mining and Technology; 2015. doi:10.13140/RG.2.2.14475.90402

(57) Wei Z, Zhen H, Fu J, Leung C, Cheung C, Huang Z. Experimental and numerical study on the laminar burning velocity of hydrogen enriched biogas mixture. *Int J Hydrog Energy*. 2019;44(39):22240–22249. doi:10.1016/j.ijhydene.2019.06.097

CHAPTER 4

V-Shaped Roughened Geometries and Their Effect on Heat Transfer and Friction Factor in Solar Air Heater

Atul Lanjewar and Sumer Singh Patel

Mechanical Engineering Department, Maulana Azad National Institute of Technology, Bhopal, India

Abstract

Solar air heater (SAH) is the easiest and the most effective way to utilize and convert solar energy into thermal energy for heating applications. SAHs are used widely in household and industrial applications. Poor thermal efficiency of SAHs has encouraged researchers to improve its thermal performance. The use of artificial roughness on absorber plate surface is the key technique for augmenting heat transfer with minimal friction factor penalty. Due to artificial roughness laminar sublayer developed below absorber plate gets broken and helps to increase turbulence of air leading to an increase in heat transfer from absorber plate to air. Numerous studies have been conducted to find the effect of various V-rib artificial roughness geometries on heat transfer and frictional performance of SAHs. In the current book chapter, an attempt has been made to summarize V-shaped artificial roughness geometries used in SAH that augments its performance. Correlations developed for heat transfer and friction factor for V-rib geometries used in SAHs by various researchers have been presented. Based on heat transfer and friction factor correlations developed by investigators an attempt has also been made to compare thermo-hydraulic performance of V-rib roughness geometries used in SAHs.

Keywords: Solar air heater, heat transfer, V-rib, solar energy

1.0   Introduction

Energy is obtainable in many different forms and plays an essential role in worldwide economic development and industrialization. With population growth, the need for energy is increasing, and achieving future energy needs with currently available energy sources is next to impossible. Exhaustion of nonrenewable energy sources in the near future is inevitable. Apart from this, nonrenewable energy sources are dangerous for life on earth. For this reason, the researchers are motivated to investigate nonconventional or alternate energy sources. Several options of renewable energy sources are available such as solar energy, wind energy, biomass energy, hydro energy, ocean and tidal energy [1]. Solar energy is the greatest promise among all these options. It is easily obtainable and there is no polluting effect on the environment. Solar energy can be used directly or indirectly by converting it into thermal energy and it is more useful if it is converted into thermal energy. Solar air heater (SAH) is a device that converts solar energy into thermal energy. The fabrication of SAH is simple and cost-effective and it is used in many applications such as space heating, cooling, and crop drying [2]. The thermal efficiency of SAH is low due to high thermal resistance that exists between the absorber surface and flowing air.

Higher thermal resistance increases the temperature of the absorber plate surface thereby causing more heat losses to the environment. The performance of SAH is low due to the presence of laminar sublayer that gets broken by creating artificial roughness on the absorber surface. The use of artificial roughness on absorber surface augments heat transfer but it also obstructs the flow and hence the pumping power requirement increases due to increased frictional losses [3–5]. Artificial roughness is provided by fixing circular wire on absorber surface. Wire is arranged in many different forms such as transverse, inclined, arc, and V-form. Circular wire that is used is known as rib element. The pressure losses get minimized if the height of the rib element is smaller than laminar sublayer. Use of rib element causes turbulence in the laminar sublayer existing adjacent to the wall without affecting the main turbulent zone in the flow. Many researchers have reported the experimental and numerical analysis of SAH duct with V-shaped and arc-shaped roughness geometries. They have reported maximum augmentation of heat transfer and its corresponding roughness parameters and have developed the correlation for heat transfer and friction factor. The aim of the current book chapter is to review studies of SAH duct with V-shaped roughness geometries as V-rib geometry is easy and convenient to fabricate on absorber plate as compared to fabricating arc rib geometry, which is tedious and laborious to fabricate on absorber plate.

2.0   Heat Transfer Augmentation Methods

Two approaches are used for augmentation of the heat transfer coefficient between the absorber surface and flowing air. In the first approach, heat transfer area is increased by extended surfaces called fin. In the second approach, convective heat transfer is increased by creating turbulence at the heat transferring surface. The method to increase the surface area involves using corrugate absorber surface or attaching fins on the underside of the absorber surface. Nevertheless, the above techniques of enhancing heat transfer area also invite an undesired resistance to the fluid flow that result in an increase in pumping power requirement. Investigators have reported an increase in heat transfer by enhancing the heat transfer coefficient using turbulence generators viz. ribs, baffles, and delta winglets on the underside of the absorber surface and also the use of such turbulence promoters does not impose severe pressure drop penalty [6]. Thus, the technique of enhancing the heat transfer coefficient by using turbulence promoters is quite appropriate and much more useful. Drastic change in heat transfer and fluid flow pattern is achieved by using turbulence promoters and are widely used as artificial roughness on the absorber surface.

3.0 Roughness Characterization and Effect on Fluid Flow

V-rib roughness geometry and its modifications have been extensively studied by various researchers as can be seen from Ref. 6. These various V-rib geometries can be understood by specifying values of roughness geometrical parameters such as rib height (e), rib pitch (p), angle of attack (α), gap width (g), gap position (d), staggered rib pitch (p'), roughness width (w), number of gaps (N_g), and staggered rib size (r). These parameters can be expressed as dimensionless roughness parameters namely relative roughness height (e/D), relative roughness pitch (p/e), relative gap width (g/e), relative gap position (d/w), relative staggered element position (p'/p), relative roughness width (W/w), and relative staggered rib size (r/e). Apart from the study of V-roughness geometries researchers have also studied the effect of duct dimension such as duct width (W) and duct height (H) expressed in dimensionless form as duct aspect ratio (W/H) on heat transfer performance. The effect of various roughness dimensionless geometrical parameters on flow pattern is given below.

3.1 Effect of Relative Roughness Height (e/D)

Relative roughness height is the ratio of rib element height to the hydraulic diameter of the duct. Creating artificial roughness on the absorber surface generates two-flow separation zones, one on each side of the rib. Flow separation is responsible for the generation of turbulence that results in augmented heat transfer and increased pressure losses. Prasad and Saini [7] reported the effect of relative roughness height (e/D) on the flow pattern as shown in Fig. 1. As relative roughness height increases the location of reattachment point moves successively toward the

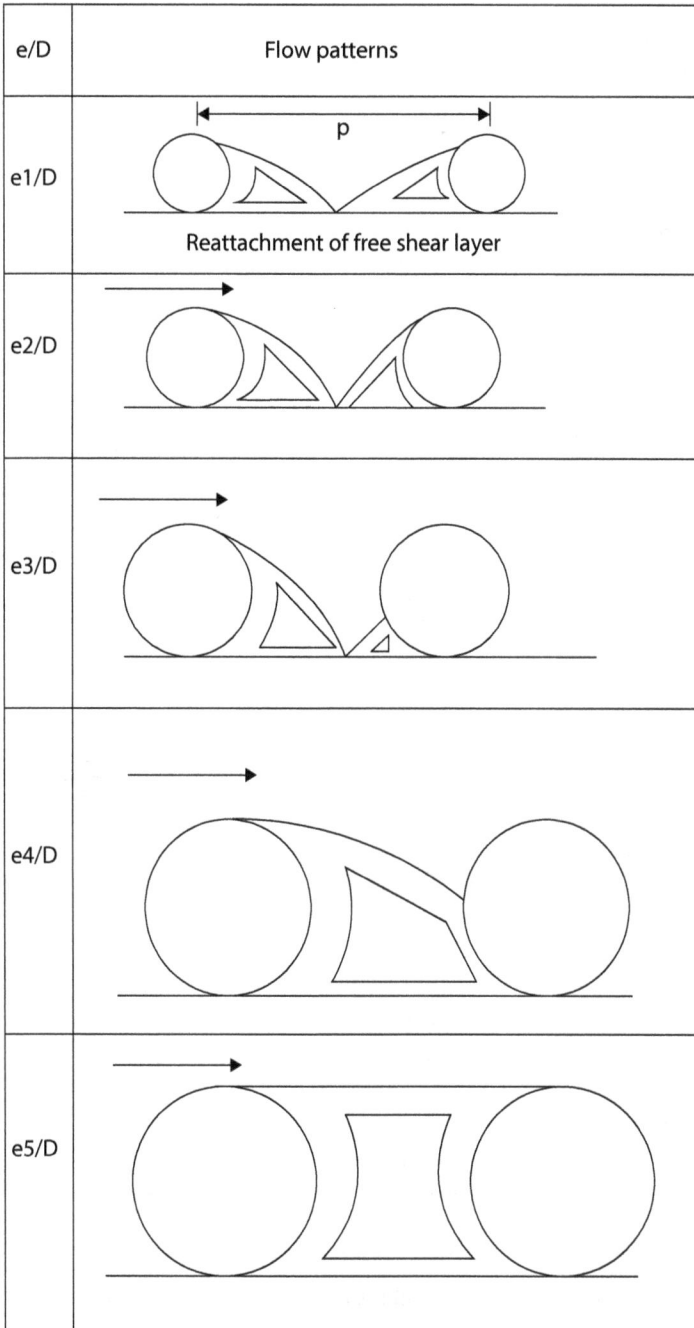

e/D	Flow patterns
e1/D	
e2/D	
e3/D	
e4/D	
e5/D	

Fig. 1: Flow Pattern as a Function of e/D [7]

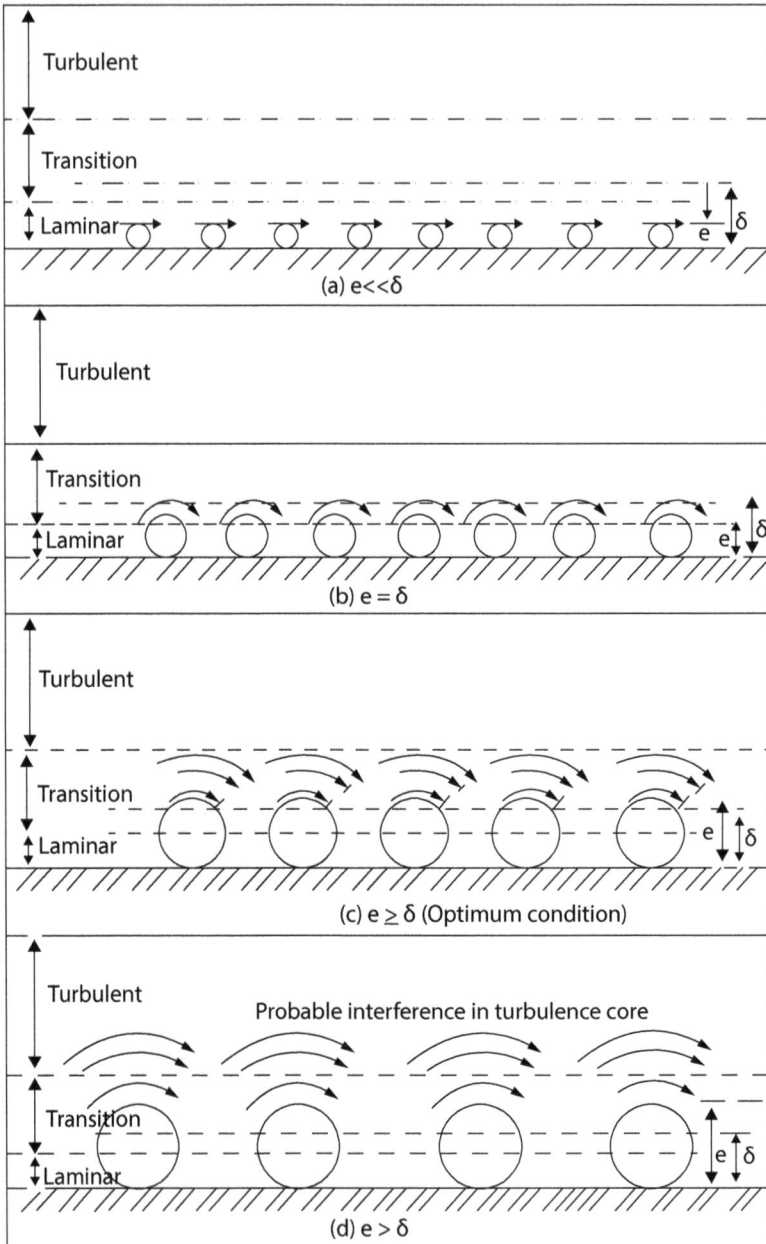

Fig. 2: Effect of Roughness Height on the Laminar Sublayer [8]

downstream rib. For large values of rib height, reattachment point does not occur and instead the flow strikes the rib resulting in poor heat transfer. As rib height is still increased, the flow passes over downstream rib with no possibility of reattachment point. Verma and Prasad [8] classified roughened surface in three categories based on rib height (e) and thickness of laminar sublayer as shown in Fig. 2.

As seen in Fig. 2, very small size rib elements remain submerged in laminar sublayer. With increase in size of rib element, they extend beyond laminar sublayer thus destroying laminar sublayer and improving heat transfer. With still further increase in the size of rib element, although the laminar sublayer gets broken the rib elements penetrate deep into turbulent core increasing frictional losses in the flow at a greater rate than increase in heat transfer. It can be summarized as

a. If $e \ll \delta$, roughness has no effect
b. If $e \gg \delta$, roughness has more effect on fluid pressure as compared to heat transfer
c. If $e \geq \delta$, the intended purpose of noticeable enhancement in heat transfer and moderate fluid pressure could be served.

3.2 *Effect of Relative Roughness Pitch (p/e)*

It is defined as the ratio of the pitch (p) to the rib element height (e) [7,8]. Many researchers performed an experiment with different pitch and reported optimum pitch based on thermal hydraulic performance (THPP) [3,4]. Prasad and Saini [7] reported the effect of relative roughness pitch (p/e) on flow pattern downstream as shown in Fig. 3. They reported that the reattachment free shear layer does not occur when relative roughness pitch is less than 8–10 due to separation at the rib. For p/e < 8, the reattachment point does not occur resulting in decrease of heat transfer and for p/e > 10, the overall number of reattachment points gets reduced because on the absorber surface the number of rib elements decreases. Another reason for low heat transfer at higher p/e is that the airflow travels longer distance after reattachment till it strikes the next rib elements and some part of the absorber surface redevelops the laminar sublayer.

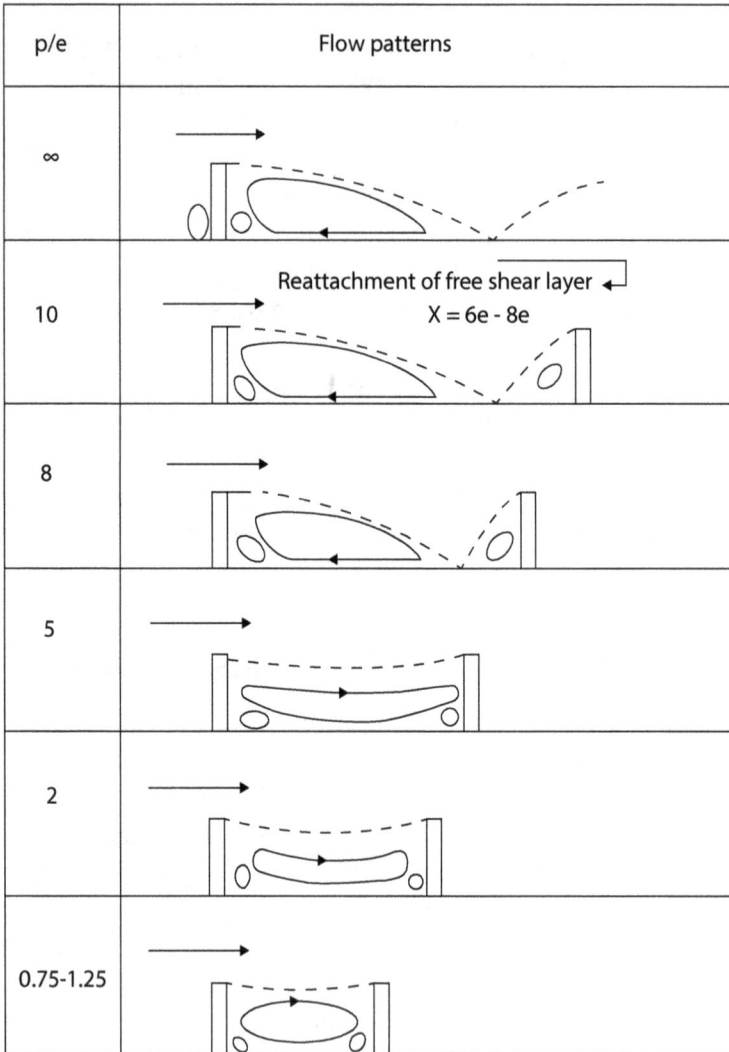

p/e	Flow patterns
∞	
10	Reattachment of free shear layer X = 6e - 8e
8	
5	
2	
0.75-1.25	

Fig. 3: Flow Pattern as Function of p/e [7]

3.3 *Effect of Flow Angle of Attack (α)*

Han et al. [9] reported the effect of four different flow angle of attack on heat transfer and friction factor. They concluded decrease in friction factor with decreasing flow angle of attack. Taslim et al. [10] performed an experiment on channel with transverse- and inclined-shaped geometries. They reported the heat transfer augmentation is higher for airflow in channel with inclined roughness as compared to transverse roughness. As per Taslim et al. [10], it is due to the generation of secondary flow cells in the case of incline rib that is responsible for higher heat transfer augmentation. These secondary flow cells also generate a spanwise variation in heat transfer. The vortices travel along the rib element and subsequently join the mainstream as shown Fig. 4. Turbulence generated by moving vortices bring in cooler channel fluid in contact with leading end thereby increasing heat transfer while the trailing end heat transfer is relatively lower due to successive heating of fluid along rib leading to spanwise variation in heat transfer. Further augmentation in heat transfer takes place by breaking the angled rib into two half ribs in the form of V-shape. From Fig. 5 in V-shape rib, there is the formation of two secondary flow cells in comparison to inclined rib and results in an overall increase in heat transfer.

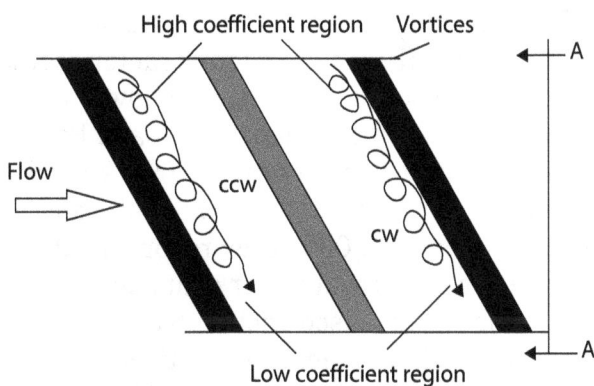

Fig. 4: Effect of Inclination of Rib on Heat Transfer [10]

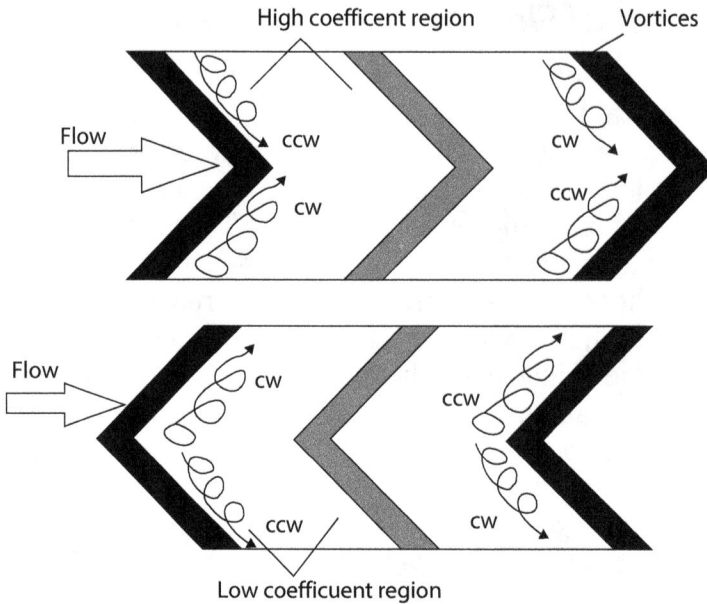

Fig. 5: V-Shape Ribs Effect on Secondary Flow Cell [10]

3.4 Effect of Duct Aspect Ratio (W/H)

It is the ratio of duct width (W) to duct depth (H). Karwa et al. [11] performed an experiment with different duct aspect ratio (W/H). They concluded a decrease in heat transfer with increase in duct aspect ratio (W/H) and the friction factor decrease with an increase in duct aspect ratio (W/H). The lower value of (W/H) gives better heat transfer performance. Han and Park [12] reported enhancement in Nusselt number (Nu) and friction factor (f) corresponding to duct aspect ratio (W/H) of 2 and 4 and Reynolds number (Re) of 30,000. They reported enhancement in Nusselt number in both rectangular channels as almost the same but there was an increased friction factor of 7 and 16 for duct aspect ratio (W/H) of 2 and 4, respectively. For the same pumping power, the square channel gives slightly better heat transfer than the rectangular duct corresponding to duct aspect ratio (W/H) of 2 and about 10% better than that of duct aspect ratio (W/H) of 4.

3.5 Effect of Relative Gap Width (g/e)

It is the ratio of gap width (g) to rib element height (e). Aharwal et al. [13] performed an experiment on inclined rib with different relative gap width (g/e). They reported the maximum and minimum augmentation in heat transfer corresponding to g/e of 1 and 2, respectively. As per Aharwal et al. [13] for g/e of 2, the flow velocities through the gap decrease due to which flow acceleration through the gap is not strong enough and hence there is a decrease in augmentation of heat transfer. For g/e less than 1, there is very little space for fluid flow through a gap that results in reducing the turbulence and hence decrease in the augmentation of heat transfer. The flow pattern of secondary flow for inclined rib with a gap is shown in Fig. 6.

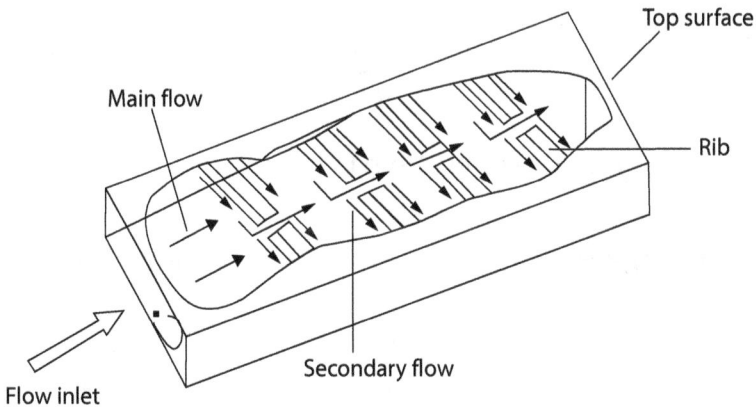

Fig. 6: Flow Pattern of Inclined Rib With Gap [13]

3.6 Effect of Relative Gap Position (d/w)

Singh et al. [14] performed experiment with different relative gap position (d/w) of V-rib as shown in Fig. 7. As stated by Singh et al. [14] providing the gap toward V-apex region corresponding to d/w of 0.65 increases heat transfer due to breaking of boundary layer thickness that progressively increases from walls to apex region. This breaking of boundary layer along with local mixing of secondary flow with main flow enhances heat transfer up to d/w of 0.65. Thereafter for the case of gaps placed very near to apex region corresponding to d/w > 0.65, there is an overlapping of two high heat transfer regions developed on gap downstream in both legs that do not cause a substantial increase in heat transfer.

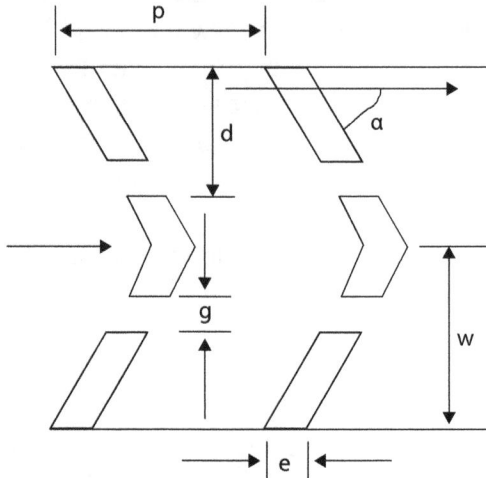

Fig. 7: Discrete V-rib [14]

3.7 *Effect of Relative Roughness Width (W/w)*

Hans et al. [15] investigated roughened SAH duct with different relative roughness width (W/w) as shown in Fig. 8. The heat transfer increases with an increase in relative roughness width (W/w) and attains maxima at (W/w) of 6 and then decreases with further increase in (W/w). As stated by Hans et al. [15] with increase in relative roughness width (W/w), the number of leading ends and secondary flow cells increase resulting in considerable augmentation in heat transfer reaching maxima for relative roughness width (W/w) of 6 on account of maximum turbulence for this value. Beyond this value, heat transfer rate reduces due to flow separation from rib top surface and causing development of boundary layer that retards heat transfer.

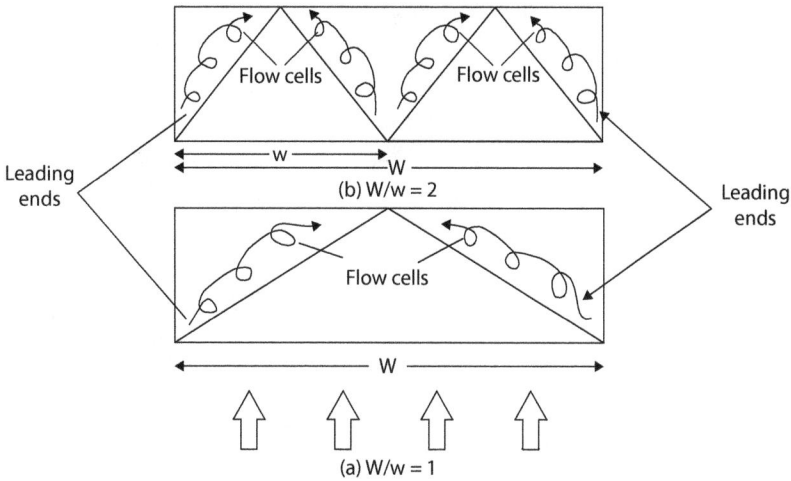

Fig. 8: Formation of Flow Cells With Increase in W/w [15]

3.8 *Effect of Relative Staggered Element Position (p'/p)*

Patil et al. [16] reported the effect of different staggered element position on heat transfer. As per Patil et al. [16], a staggered element placed in the path of flow in front of a gap results in flow separation, scattering, and reattachment in the interrib region accompanied by low heat transfer zone in the upstream and downstream of staggered element. Location of flow separation and reattachment depends on the position of staggered element implies that the staggered element placed just after the halfway of roughness pitch separates the flow and restarts the boundary layer after extracting maximum energy from the surface that augments heat transfer. On the other hand, staggered element placed close to downstream of gap leads to early flow separation, while the flow separation is delayed if it is placed close to upstream of gap that reduces heat transfer in these both cases. Figure 9 shows the flow pattern of discrete V-rib combined with staggered element.

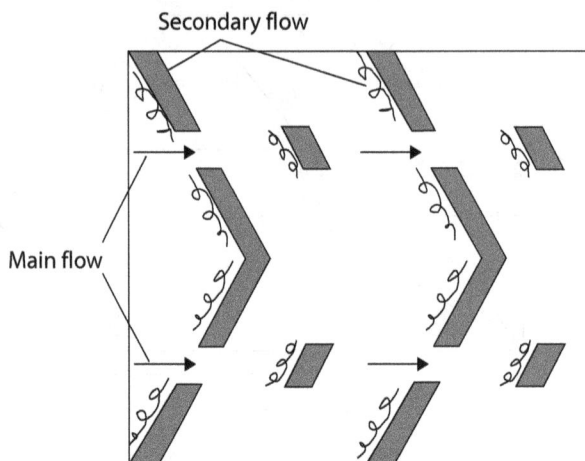

Fig. 9: Flow Pattern of Broken V-Shaped Combined With Staggered Element [16]

3.9 *Effect of Relative Staggered Size (r/e)*

Patel and Lanjewar [17] also investigated the effect of different staggered element size on heat transfer. When flow strikes the staggered element, it scatters on the two sides of the staggered element. The vortices interfere with the primary stream that produces reasonable fluid mixing mainly on the two sides of the staggered element. As the length of staggered element increases, the region affected by the disturbance due to interference on the two sides broaden, which leads to an acceleration in turbulence near the staggered element region corresponding to r/e of 3.5. Beyond this length of staggered element, the flow that scatters due to staggered element interferes with the reattachment point of the mainstream leading to a decrease in heat transfer for an increase in relative staggered element size beyond 3.5.

4.0 Effect of V-Shaped Roughness on Heat Transfer and Friction Factor

Momin et al. [18] performed experiment of SAH duct with V-rib having geometrical parameters such as e/D of 0.02–0.034, α of 30°–90°, p/e of 10, and Re of 2,500–18,000 as shown in Fig. 10.

They reported that the rate of increment of heat transfer decreases with an increase in Reynolds number whereas pressure loss increases. It is due to the nonreattachment of free shear layer at a higher value of e/D due to which the rate of augmentation of heat transfer is not proportional to that of pressure loss. Maximum augmentation of heat transfer and pressure loss is 2.30 and 2.83 times, respectively, as compared to smooth surface corresponding to angle of attack (α) of 60° and the maximum THPP is also obtained at angle of attack (α) of 60°. Increase in heat transfer is due to roughness that causes reattachment, separation of flow, and secondary flow generation. They also reported that V-rib has better performance as compared to inclined rib roughness due to the generation of more secondary flow cells.

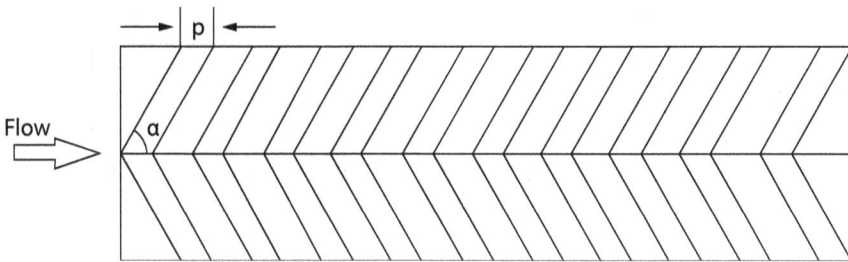

Fig. 10: Continuous V-rib Geometry of Solar Air Heater [18]

Karwa [19] investigated transverse, inclined, V-continuous, and V-discrete roughness with duct aspect ratio (W/H) from 7.19 to 7.75, e/D of 0.0467–0.05, p/e of 10, α of 60°, and relative roughness length for discrete roughness (B/S) of 3. On the basis of experimental investigation, he reported V-discrete geometry gives the best performance as compared to other geometries studied and also V-rib pointing downstream performs better than V-rib pointing upstream.

Singh et al. [20] performed an experiment of SAH duct with roughened surface having V-rib with gap roughness with p/e of 4–12, e/D of 0.015–0.043, α of 30°–75°, g/e of 0.5–2, d/w of 0.2–0.8, and Re of 3,000–15,000 as shown in Fig. 7. They reported that V-rib with gap gives maximum heat transfer as compared to continuous V-rib. As per Singh et al. [20] providing a gap in continuous V-rib augments, the local heat transfer coefficient at the downstream of the gap thereby causing an enhancement in average Nu. This local heat transfer coefficient augmentation is caused by enhanced flow mixing and turbulence resulting from accelerated flow through the gap. They also reported the maximum augmentation of heat transfer and friction factor is obtained corresponding to p/e of 8, e/D of 0.043, α of 60°, g/e of 1 and d/w of 0.6, and developed correlation for heat transfer and friction factor.

Maithani and Saini [21] performed experiment of SAH duct having V-shaped with symmetrical gaps as shown in Fig. 11. They investigated the effect of p/e of 6–12, α of 30°–75°, g/e of 1–5, N_g of 1–5, e/D of 0.043, and Re of 4,000–18,000 on heat transfer and friction factor. They reported maximum augmentation of heat transfer and friction factor as 3.6 and 3.67 times, respectively, as compared to smooth surface corresponding to p/e of 10, α of 60°, g/e of 4, N_g of 3, and e/D of 0.043. They also reported that heat transfer increased with an increasing number of gaps up to 3, and after this value, it decreased because the number of gaps increases the open area thus reducing turbulence intensity.

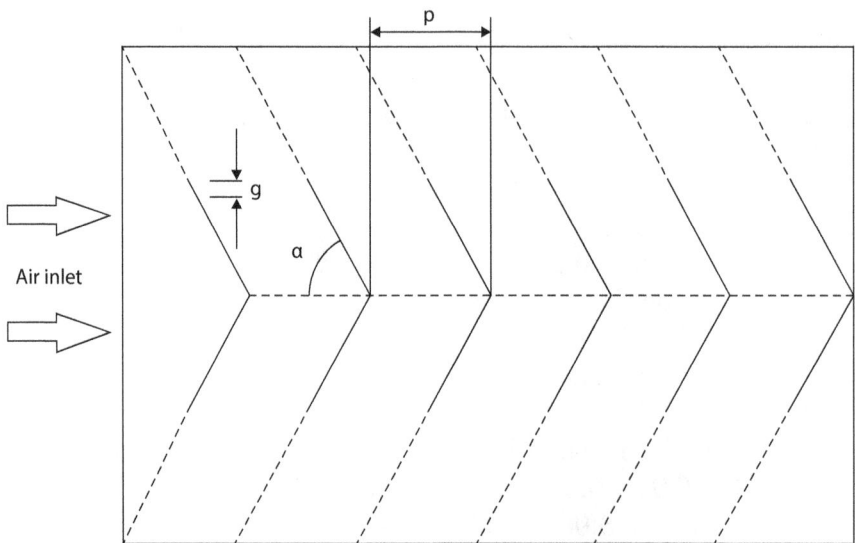

Fig. 11: V-rib With Symmetrical Gaps Geometry of Solar Air Heater [21]

Karwa and Chitoshiya [22] performed an experiment of
SAH duct with discrete V-down rib having p/e of 10.63, e/D
of 0.047, α of 60°, W/H of 7.8, relative roughness width of
2.06, relative roughness length ratio (B/S) of 6, and Re of
2,750–11,150 as shown in Fig. 12 in outdoor condition. As per
Karwa and Chitoshiya [22] by using discrete ribs, there is better
mixing of secondary flow with primary airflow as secondary
flow mixes with primary flow only after short distance because
of rib discretization. This effect increases with an increase in
discretization level resulting in increased turbulence because of
better mixing of primary and secondary flow. They reported that
maximum augmentation in thermal efficiency is 12.5%–20%.
Karwa and Chauhan [23] developed mathematical model of SAH
duct with roughened surface for the analysis of different operating
and design parameters such as p/e of 10, e/D of 0.02–0.07, B/S
of 6, relative roughness width (w/e) of 2, e^+ of 15–75, collector
slope (β) of 0° horizontal and 45°, G of 0.010–0.06 kg s^{-1}m^{-2}, and
Re of 1,070–26,350 on the thermal and effective efficiency. They
reported maximum augmentation in thermal efficiency is obtained
at G $\leq$ 0.04 kg s^{-1}m^{-2} and at higher flow rate effective efficiency is
considerably lower as compared to thermal efficiency. It is due to
increase in pumping power with increase in flow rate.

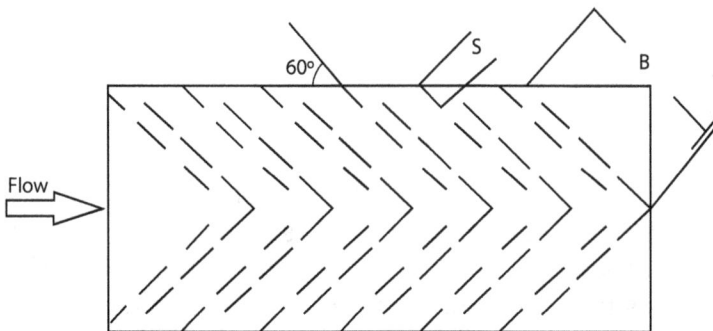

Fig. 12: V-Down Discrete Rib Roughness Geometry of
Solar Air Heater [22]

Hans et al. [15] performed experiment of SAH duct with
roughened surface having multi V-shaped roughness having p/e
of 6–12, e/D of 0.019–0.043, α of 30°–75°, W/w of 1–10, and Re

of 2,000–20,000 as shown in Fig. 13. They reported maximum augmentation of heat transfer and friction factor as 6 and 5 times, respectively, as compared to smooth surface corresponding to p/e of 8, W/w of 6 and 10 and α of 60°. The increase in Nusselt number as per Hans et al. [15] is due to an increase in the number of leading ends and secondary flow cells of V-rib as relative roughness width is increased. Maximum increase in Nusselt number occurs at relative roughness width of 6 as for this value turbulence achieved is highest and further increase in relative roughness width results in flow separation from the top of rib surface and boundary layer redevelopment, whereas friction factor keeps on increasing due to formation of vortices caused by flow separation. They also developed a correlation for heat transfer and friction factor.

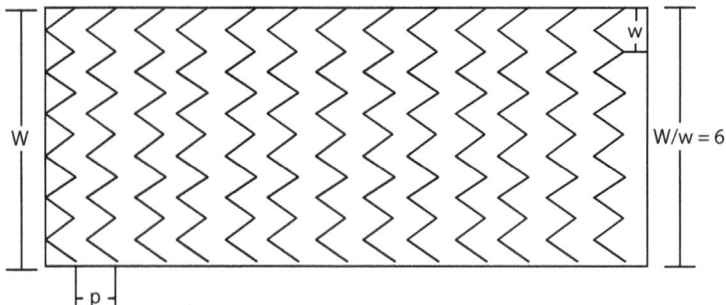

Fig. 13: Multi V-Shaped Roughness Geometry of Solar Air Heater [15]

Kumar et al. [24] performed an experiment of SAH duct with multi V-shaped roughness with a gap as shown in Fig. 14. They performed experiments with different combinations of roughness parameters such as W/H of 12, p/e of 6–12, e/D of 0.022–0.043, α of 30°–75°, g/e of 0.5–1.5, W/w of 1–10, relative gap distance (G_d/L_v) of 0.24–0.80 and Re of 2000–20,000. Kumar et al. [24] reported that the provision of a gap in multi V-shaped roughness allows the release of secondary flow and it mixes with the main flow through the gap as shown in Fig. 14. The main flow passing through the gap is developed flow with a thicker boundary layer consisting of a viscous sublayer. Due to gap presence secondary flow along the rib joins the main flow to accelerate it which energizes the retarded boundary layer flow along the surface resulting in enhancement of heat transfer through the gap width area behind the ribs. They also reported maximum augmentation of heat transfer and friction factor as 6.74 and 6.37 times, respectively, as compared to smooth surface corresponding to p/e of 8, e/D of 0.043, α of 60°, g/e of 1, relative gap distance (G_d/L_v) of 0.69, and W/w of 6 and 10 for heat transfer and friction factor respectively.

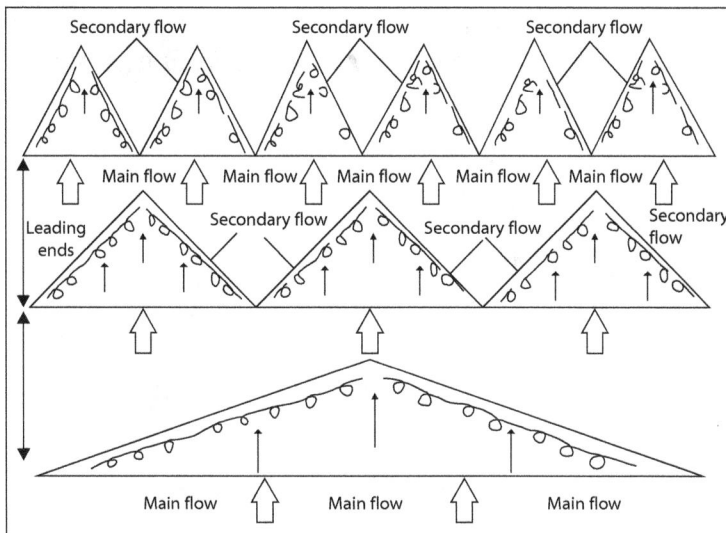

Fig. 14: Multi V-Shaped Rib With Gap Geometry of Solar Air Heater [24]

Patil et al. [25] performed an experiment of SAH duct with broken V-rib combined with staggered element having height e/D of 0.043, g/e of 1, α of 60°, d/w of 0.2–0.8, p'/p of 0.2–0.8, r/e of 1–2.5, W/H of 12 and Re of 3000–17,000 as shown in Fig. 15. They reported that the roughened surface prepared with d/w of 0.6, p'/p of 0.6, and r/e of 2.5 gives the maximum heat transfer and friction factor. As per Patil et al. [25] for gap located near the leading edge of V-rib leads to the release of secondary flow before it achieves saturation while it gets heated moving along V-rib. Also, secondary flow strength is not that sufficient to accelerate the main flow through the gap, whereas for the other case when the gap is located very near to V-apex merging of two secondary flows effectively reduces their strength leading to lower heat transfer. As per their study, they reported an optimum location of gap (d/w) as 0.6. As per Patil et al. [25] staggered rib placed at (p'/p) of 0.6 removes maximum thermal energy from absorber surface leading to heat transfer enhancement. For staggered rib placed very close to gap results in early separation of flow and for staggered rib placed much after gap corresponding to (p'/p) of 0.8 results in delay of flow separation. With an increase in staggered rib size, thermal performance improves attains maxima at r/e of 2.5 thereafter performance declines due to rise in frictional losses because of increased turbulence.

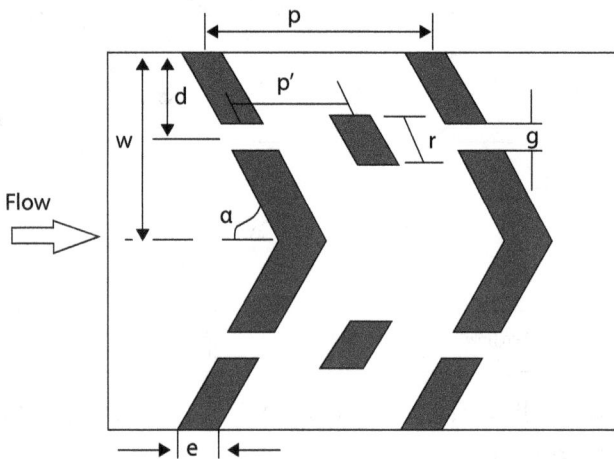

Fig. 15: Broken V-Shaped Combined With Staggered Element Geometry of SAH [25]

Deo et al. [26] performed an experiment of SAH duct with multigap V-rib combined with staggered element having p/e of 4–14, e/D of 0.026–0.057, α of 40°–80°, g/e of 1, r/e of 4.5, p'/p of 0.65, no of gaps on each side of V-rib (N_g) of 2, W/H of 12, and Re of 4000–12,000. As per Deo et al. [26], flow patterns responsible for the augmentation of heat transfer are (a) reattachment of the primary flow behind the V-rib, Fig. 16(a) and 16(b), (b) reattachment of the primary flow behind the staggered element, Fig. 16(c) (for higher pitch value), (c) movement of the secondary flow along the V-rib and its refreshment through the gap, (4) acceleration of the primary flow through the gaps, (5) generation of the counter-rotating secondary flow vortices across the cross-section of the duct, Fig. 16(d), (6) generation of additional turbulence due to scattering of the primary flow from the staggered element, Fig. 16(a), (7) faster mixing of the flow in the turbulence wake region, Fig. 16(a) of the staggered element (for higher pitch value). They reported maximum augmentation of heat transfer and THPP as 3.34 and 2.45 times, respectively, as compared to smooth surface corresponding to p/e of 12, e/D of 0.044, α of 60°, and Re of 12,000; whereas maximum increase in friction factor as 3.38 times as compared to smooth surface for p/e of 12, e/D of 0.044, α of 80°, and Re of 12,000.

Ravi and Saini [27] investigated discrete multi V-shaped and staggered rib roughness as an artificial roughness with single-pass as well as double-pass SAH. The roughness geometry is shown in Fig. 17. The experimental study involved the range of Reynolds number from 2,000 to 20,000 and relative staggered rib size (r/e) from 1 to 4, relative staggered rib pitch (p'/p) from 0.2 to 0.8, and relative roughness width (W/w) from 5 to 8 with other parameters were kept constant with angle of attack (α) as 60°, relative gap distance (G_d/L_v) as 0.70, relative pitch ratio (P/e) as 10, e/D$_h$ as 0.043 and g/e as 1. They reported maximum Nusselt number corresponding to r/e = 3.5, p'/p = 0.6, and W/w = 7.

Jain and Lanjewar [28] performed an experiment of SAH duct with V-rib with symmetrical gap and staggered element geometry having p/e of 10–16 as shown in Fig. 18. Jain and Lanjewar [28] reported maximum augmentation of heat transfer and friction factor as 2.30 and 3.18 times, respectively, as compared to smooth surface corresponding to p/e of 12. They have shown experimentally that V-rib with symmetrical gap and staggered element geometry is thermo-hydraulically better than V-rib with symmetrical gaps geometry of Maithani and Saini [21] and also better than multigap V-rib combined with staggered element geometry of Deo et al. [26]

Patel and Lanjewar [29] modified geometry of Jain and Lanjewar [28] by introducing gap in the rib element of V-rib as

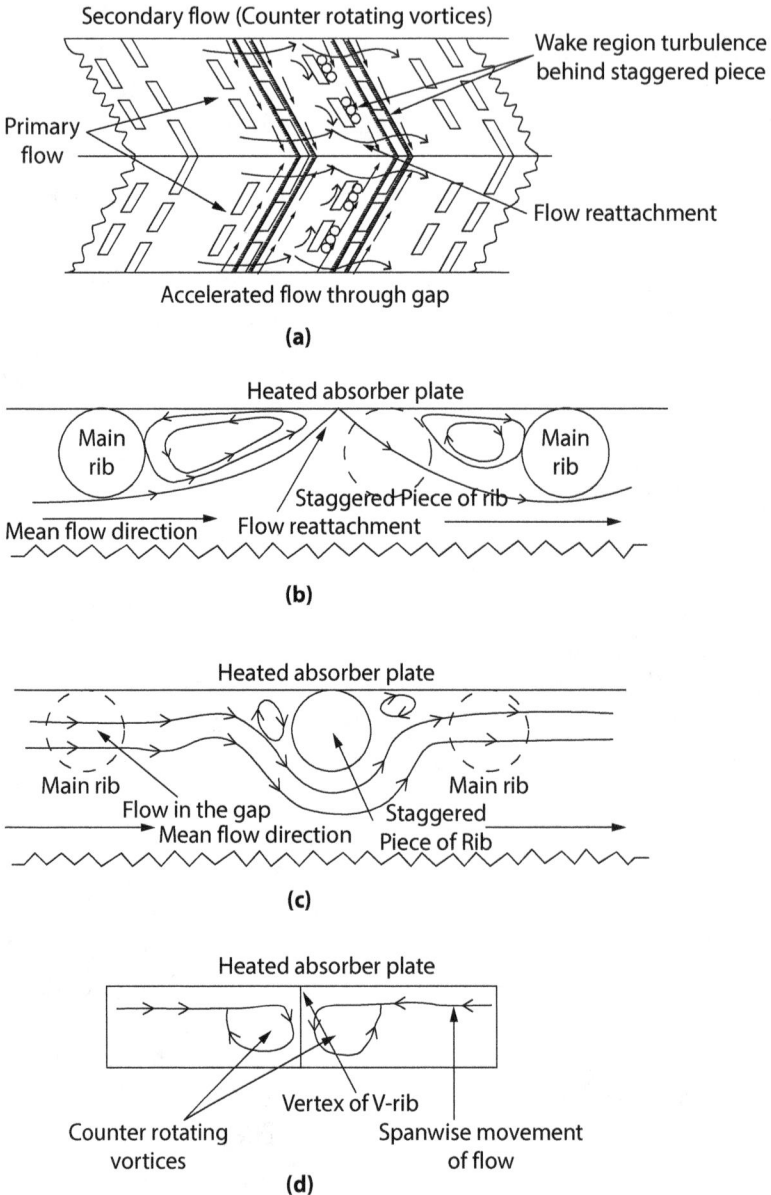

Secondary flow (Counter rotating vortices)

Wake region turbulence behind staggered piece

Primary flow

Flow reattachment

Accelerated flow through gap

(a)

Heated absorber plate

Main rib

Main rib

Staggered Piece of rib

Mean flow direction   Flow reattachment

(b)

Heated absorber plate

Main rib

Main rib

Flow in the gap   Staggered

Mean flow direction   Piece of Rib

(c)

Heated absorber plate

Vertex of V-rib

Counter rotating vortices

Spanwise movement of flow

(d)

Fig. 16: (a) Flow Patterns at Various Locations Between the Upstream and Downstream ribs. (b) Nearer Wall Flow Between Two Adjacent V-Ribs. (c) Near Wall Flow Coming Through the Gap and Moving Over the Staggered Piece of Rib. (d) Movement of the Counter-Rotating Vortices in the Plane of Cross-Section of the Duct Due to the Presence of Rib [26]

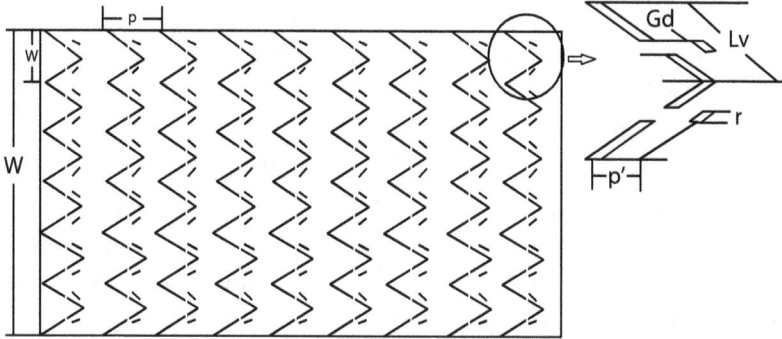

Fig. 17: Discrete Multi V-Shaped and Staggered Rib Roughness Geometry of SAH [27]

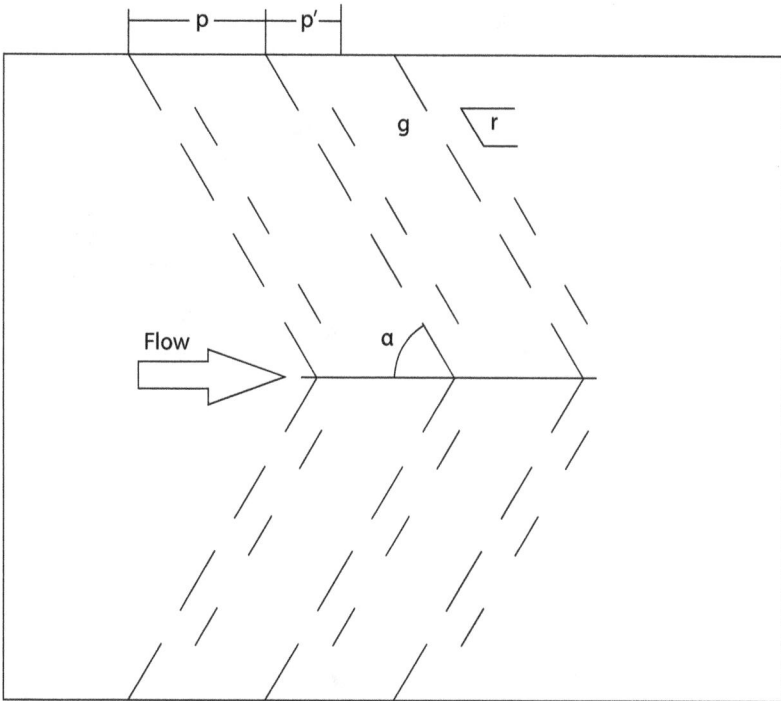

Fig. 18: V-rib With Symmetrical Gap and Staggered Element Geometry of Solar Air Heater [28]

shown in Fig. 19. They studied variation of relative roughness pitch on heat transfer and friction factor and reported best results for relative roughness pitch of 10. This geometry has been reported to also perform significantly better than multigap V-rib combined with staggered element geometry of Deo et al. [26]

The optimum value of roughness parameters reported by various investigators for V-rib roughness is presented in Table 1. Table 2 presents the correlation of heat transfer and friction factor for different V-rib roughness geometries used in SAH.

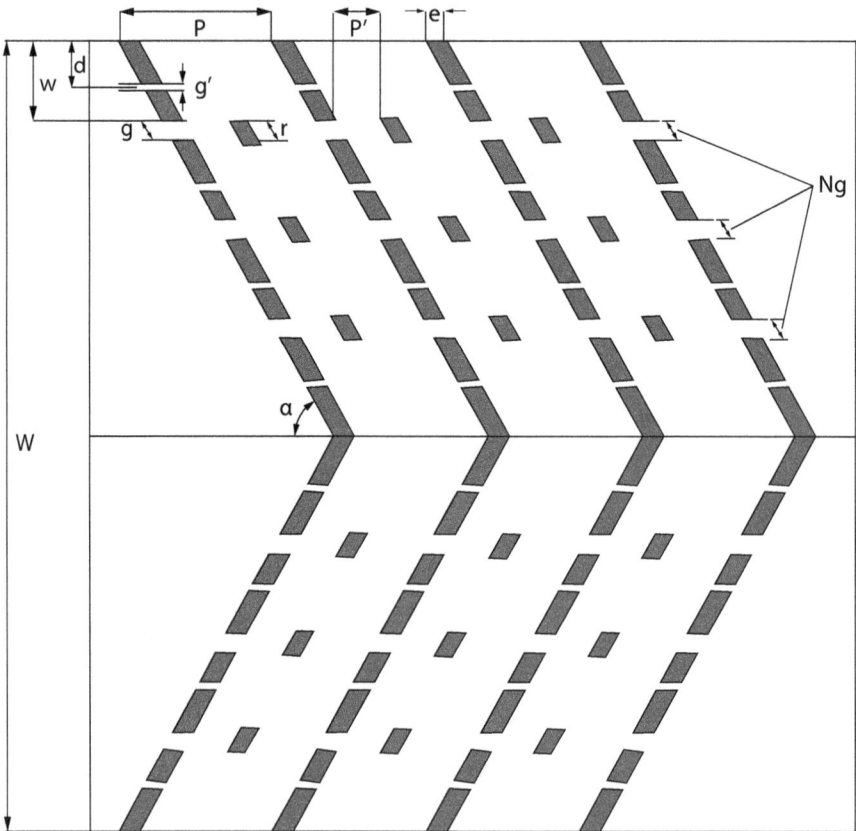

Fig. 19: Gap in V-rib With Symmetrical Gap and Staggered Element Geometry of SAH [29]

Table 1: Roughness Geometry With Optimized Details

S. No.	Authors	Roughness Elements	Optimum Values of Parameters			
			p/e	e/D_h or e/D	α	Others
1.	Momin et al. [18]	Continuous V-ribs	10	0.034	60°	
2.	Karwa [19]	V-down discrete roughness	10	0.050	60°	
3.	Singh et al. [20]	V-ribs with gap	8	0.043	60°	d/w = 0.65; g/e = 1; W/H = 12
4.	Maithani and Saini [21]	V-ribs with symmetrical gaps	10	0.043	60°	g/e = 4; N_g = 3
5.	Karwa and Chitoshiya [22]	Discrete V-down rib	10.63	0.047	60°	B/S = 6; W/w = 2.06; W/H = 7.8
6.	Karwa and Chauhan [23]	Discrete V-down rib	10	0.07	60°	B/S = 6; W/w = 2; e^+ = 15–75
7.	Hans et al. [15]	Multi V-shaped ribs	8	0.043	60°	W/w = 6 (For Nu) and 10 (For f)

(Continued)

Table 1 : Roughness Geometry With Optimized Details (*Continued*)

S. No.	Authors	Roughness Elements	Optimum Values of Parameters			
			p/e	e/D_h or e/D	α	Others
8.	Kumar et al. [24]	Multi V-shaped roughness with gap	8	0.043	60°	W/w = 6 (For Nu) and 10 (For f); G_d/L_v = 0.69; g/e = 1; W/H = 12
9.	Patil et al. [25]	Broken V-rib combined with staggered element	10	0.043	60°	d/w = 0.6; g/e = 1; p'/p = 0.6; r/e = 2.5 W/H = 12
10.	Deo et al. [26]	Multigap V-rib combined with staggered element	12	0.044	60°	g/e = 1; p'/p = 0.65; r/e = 4.5; W/H = 12
11.	Ravi and Saini [27]	Discrete multi V-shaped and staggered rib roughness	10	0.043	60°	g/e = 1, W/w = 7, r/e = 3.5, p'/p = 0.6, G_d/L_v =0.7
12.	Jain and Lanjewar [28]	V-rib with symmetrical gap and staggered element	12	0.045	60°	N_g = 3; g/e = 4; p'/p = 0.65; r/e = 4; W/H = 8

Table 2: Correlations Developed for Different V-rib Geometries Used in SAH

Authors	Roughness Geometry	Reynolds Number Range	Heat Transfer Correlation	Friction Factor Correlation
Momin et al. [18]	Continuous V-ribs	Re = 2,500–18,000	$Nu = 0.067\,(e/D_h)^{0.424}\,Re^{0.888}\,(\alpha/60)^{-0.077}\,[exp\{-0.782\,(ln\,(\alpha/60))^2\}]$	$f = 6.266\,(e/D_h)^{0.565}\,Re^{-0.425}\,(\alpha/60)^{-0.093}\,[exp\{-0.719\,(ln\,(\alpha/60))^2\}]$
Karwa [19]	V-down discrete roughness	Re = 2,800–15,000	$G=[(f/2St)-1]/(2f)^{0.5}+R$	$R = (2f)^{0.5} + 2.5ln(2e/D) + 3.75$ $e^+=(2f)^{0.5} \times Re \times (e/D)$
Singh et al. [20]	V-ribs with gap	Re = 3,000–15,000	$Nu = 2.36 \times 10^{-3}\,Re^{0.90}\,(p/e)^{3.50}\,(\alpha/60)^{-0.023}\,(d/W)^{-0.043}\,(g/e)^{-0.014}\,(e/D_h)^{0.47}\,exp(-0.84\{ln\,(\alpha/60)^2\})$ $exp(-0.72\{ln\,(\alpha/60)^2\})$ $exp(-0.05\{ln\,(d/W)^2\})$ $exp(-0.15\{ln\,(g/e)^2\})$	$f = 4.13 \times 10^{-2}\,Re^{-0.126}\,(p/e)^{2.74}\,(\alpha/60)^{-0.034}\,(d/W)^{-0.058}\,(g/e)^{0.031}\,(e/D_h)^{0.70}$ $exp(-0.685\{ln\,(p/e)^2\})$ $exp(-0.93\{ln\,(\alpha/60)^2\})$ $exp(-0.058\{ln\,(d/W)^2\})$ $exp(-0.21\{ln\,(g/e)^2\})$
Maithani and Saini [21]	V-ribs with symmetrical gaps	Re = 4,000–18,000	$Nu = 1.8 \times 10^{-5}\,Re^{0.9635}N_g^{0.126}$ $(p/e)^{5.7419}\,(\alpha/60)^{0.1307}$ $(g/e)^{0.111}exp(-1.299\{ln\,(P/e)^2\})$ $exp(-0.895\{ln\,(\alpha/60)^2\})$ $exp(-0.055\{ln\,(N_g)^2\})$ $exp(-0.0401\{ln\,(g/e)^2\})$	$f = 3.6 \times 10^{-6}\,Re^{0.1512}N_g^{0.1484}\,(p/e)^{9.24}$ $(\alpha/60)^{0.07}\,(g/e)^{0.072}exp(-2.08\{ln\,(P/e)^2\})$ $exp(-0.3364\{ln\,(\alpha/60)^2\})$ $exp(-0.0763\{ln\,(N_g)^2\})\,exp(-0.0249\{ln\,(g/e)^2\})$

(Continued)

Table 2 : Correlations Developed for Different V-rib Geometries Used in SAH (*Continued*)

Authors	Roughness Geometry	Reynolds Number Range	Heat Transfer Correlation	Friction Factor Correlation
Hans et al. [15]	Multi V-shaped ribs	Re = 2,000–20,000	$Nu = 3.35 \times 10^{-5} Re^{0.92} (e/D_h)^{0.77} (W/w)^{0.43} (\alpha/90)^{-0.49} \times exp[-0.1177(ln(W/w))^2] \times exp[-0.61(ln(\alpha/90))^2](p/e)^{8.54} \times exp[-2.0407(ln(p/e))^2]$	$f = 4.47 \times 10^{-4} Re^{-0.3188} (e/D_h)^{0.73} (W/w)^{0.22} (\alpha/90)^{-0.39} exp[-0.52(ln(\alpha/90))^2] \times (p/e)^{8.9} exp[-2.133(ln(P/e))^2]$
Kumar et al. [24]	Multi V- rib with gap	Re = 2,000–20,000	$Nu = 8.532 \times 10^{-3} (Re)^{0.932} (e/D_h)^{0.175} (W/w)^{0.506} \times exp[-0.0753(Ln(W/w))^2](G_d/L_v)^{-0.0348} \times exp[-0.0653(Ln(G_d/L_v))^2](g/e)^{-0.0708} \times exp[-0.223(Ln(g/e))^2](\alpha/60)^{-0.0239} \times exp[0.1153(Ln(\alpha/60))^2](p/e)^{1.196} \times exp[-0.2805(Ln(p/e))^2]$	$f = 3.1934 \times (Re)^{-0.3151} (e/D_h)^{0.268} (W/w)^{0.1132} \times exp[0.0974(Ln(W/w))^2](G_d/L_v)^{0.0610} \times exp[-0.1065(Ln(G_d/L_v))^2](g/e)^{-0.1769} \times exp[-0.6349(Ln(g/e))^2](\alpha/60)^{0.1553} \times exp[-0.1527(Ln(\alpha/60))^2](p/e)^{-0.7941} \times exp[0.1486(Ln(p/e))^2]$
Patil et al. [25]	Broken V-down rib combined with staggered rib	Re = 3,000–17,000	$Nu = 0.0089 \times Re^{0.97} exp[0.12/1 + \{2.42ln(d/w) + 1.19\}^2 + 0.11/1 + \{2.5ln(p'/p) + 1.41\}^2] + 0.14[ln(r/g)]^{0.71}$	$f = 0.09 \times Re^{-0.18} exp[0.10/1 + \{3.18ln(d/w) + 1.56\}^2 + 0.08/1 + \{2.6ln(p'/p) + 1.41\}^2 + 0.17[ln(r/g)]^{2.5}]$

Authors	Roughness Geometry	Reynolds Number Range	Heat Transfer Correlation	Friction Factor Correlation
Deo et al. [26]	Multigap V-down ribs combined with staggered ribs	Re = 4,000–12,000	$Nu = 0.02253 \times Re^{0.98} \times (p/e)^{-0.06} \times (e/D_h)^{0.04} \times (\alpha/60)^{0.04}$	$f = 0.3715 \times Re^{-0.15} \times (p/e)^{0.21} \times (e/D_h)^{0.65} \times (\alpha/60)^{0.57}$
Ravi and Saini [27]	Discrete multi V-shaped and staggered rib roughness	Re = 2,000–20,000	$Nu = 3.382 \times 10^{-6}\, Re^{0.9072} (p'/p)^{-0.0599} (r/e)^{0.255} (W/w)^{10.17} exp(-0.0613(\ln (p'/p))^2) exp(-0.0846(\ln (r/e))^2) exp(-2.677(\ln (W/w))^2)$	$f = 0.2698\, Re^{-0.3152} (p'/p)^{-0.0754} (r/e)^{0.1522} (W/w)^{1.6948} exp(-0.0692(\ln (p'/p))^2) exp(0.0126(\ln (r/e))^2) exp(-0.3191(\ln (W/w))^2)$

5.0 Computational Fluid Dynamics Analysis of SAH

Computational fluid dynamics (CFD) is another efficient approach to solve the problem of fluid flow and heat transfer of SAH with roughened surface. Some of the investigators have reported the CFD analysis of SAH duct with artificial roughness in V-shaped roughness pattern.

Rana et al. [30] conducted a CFD analysis of roughened SAH having V-shape rib with symmetrical gaps. The roughness parameters for analysis are p/e of 6–12, α of 30°–75°, e/D of 0.042, and Re of 3,800–18,000. They reported that maximum performance of SAH with roughened surface is obtained corresponding to p/e of 10 and α of 60°.

Jin et al. [31] conducted CFD study to investigate thermal-hydraulic performance of multi V-shaped ribs roughness in a SAH duct using the ANSYS FLUENT code and Renormalization-group k-ϵ turbulence model. They concluded that maximum average Nusselt number and thermo-hydraulic performance parameter occur at W/w of 6 for Reynolds number ranging from 8,000 to 15,000 and W/w of 3 for a Reynolds number between 18,000 and 20,000. The friction factor attains a maximum value corresponding to W/w of 6 for a Reynolds number ranging from 8,000 to 12,000 and W/w of 8 for a Reynolds number between 15,000 and 20,000.

Kumar and Kim [32] studied a CFD model of fluid flow and heat transfer in a rib roughened SAH duct using the renormalization k-ϵ model. They concluded substantial performance augmentation by using artificial discrete multi V-shape in an SAH duct and thermo-hydraulic performance with a maximum value of 3.6 was reported for roughness geometry corresponding to g/e of 1.0.

Sharma and Thakur [33] presented CFD analysis for heat transfer and friction loss characteristics in a SAH having V-shaped rib roughness at 60° angle of attack relative to flow direction pointing downstream on the underside of the absorber plate. The roughness parameters for CFD analysis were e/D_h of 0.0216–0.043 and p/e of 6–12. They reported that Nusselt number and friction factor increase with an increase in e/D_h and gives the opposite trend with an increase in p/e.

Karanth et al. [34] carried out CFD studies for arc and V-shaped ribs. They reported 1.32 times increase in Nusselt number for reverse V-ribs and 1.27 times increase in Nusselt number for reverse arc ribs as compared to smooth plate.

Dongxu et al. [35] numerically studied heat transfer characteristics of SAH using multi V-ribs. They optimized number of ribs and reported optimum pitch of 10 giving highest thermo-hydraulic performance.

6.0   Performance of V-rib Geometries

Based on correlations of Nusselt number and friction factor reported by various investigators, Fig. 20 shows variation of Nusselt number with Reynolds number for different V-rib geometries and Fig. 21 shows variation of Nusselt number with Reynolds number for different multi V-rib geometries. Among V-rib geometries considered as per Fig. 20, the V-shaped with symmetrical gaps geometry of Maithani et al. [21] gives highest Nusselt number and for multi V-rib geometries considered as per Fig. 21, the discrete multi V-shaped and staggered rib roughness of Ravi and Saini [27] gives highest Nusselt number. Similarly Fig. 22 shows a variation of friction factor with Reynolds number for different V-rib geometries and Fig. 23 shows a variation of friction factor with Reynolds number for different multi V-rib geometries.

In SAH, heat transfer enhancement is accompanied by increase in friction. For simultaneous consideration of thermal and hydraulic performance as per Lewis [36], thermo-hydraulic performance parameter can be used and is defined as $(Nu/Nu_s)/(f/f_s)^{1/3}$.

Thermo-hydraulic performance parameter has been plotted for V-rib and multi V-rib geometries of SAH. Figure 24 shows variation of thermo-hydraulic performance parameter with Reynolds number for different V-rib geometries. Figure 25 shows the variation of thermo-hydraulic performance parameter with Reynolds number for different multi V-rib geometries. Among V-rib

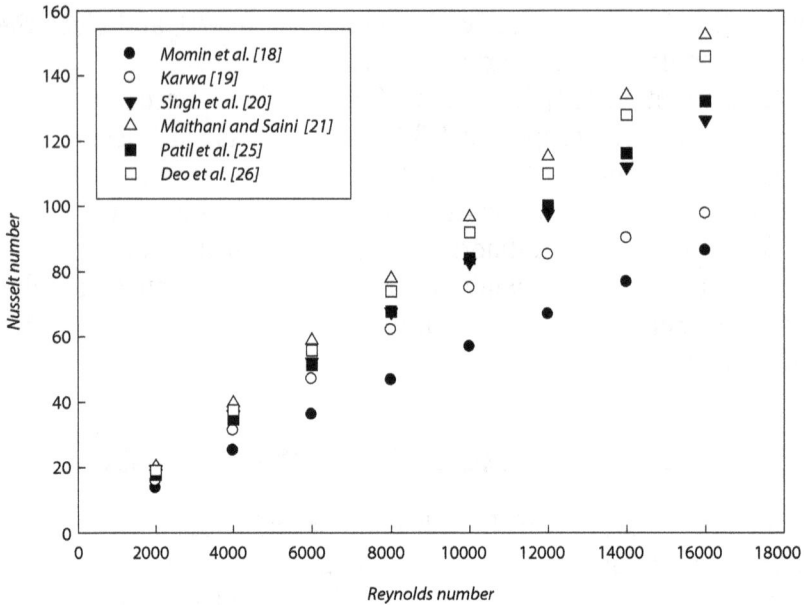

Fig. 20: Variation of Nusselt Number With Reynolds Number for Different V-rib Roughness Geometries in SAH

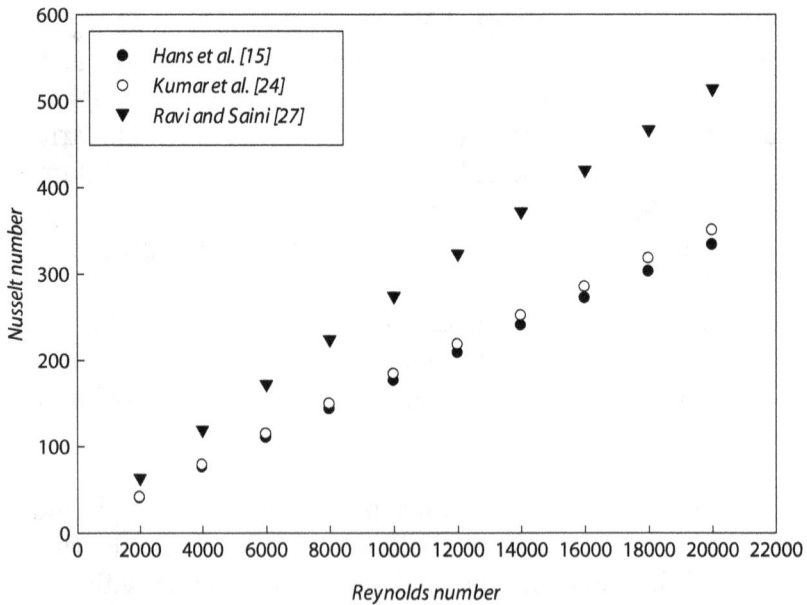

Fig. 21: Variation of Nusselt Number With Reynolds Number for Different Multi V-rib Roughness Geometries in SAH

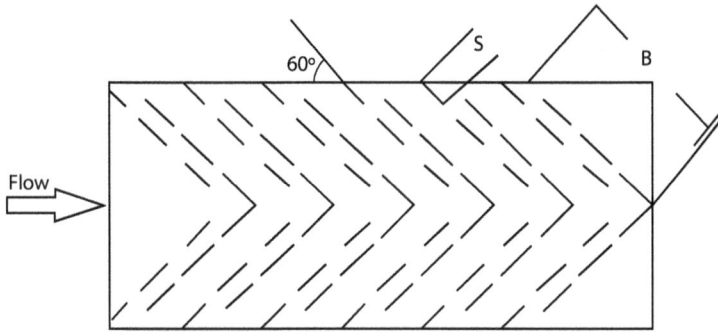

Fig. 22: Variation of Friction Factor With Reynolds Number for Different V-Rib Roughness Geometries in SAH

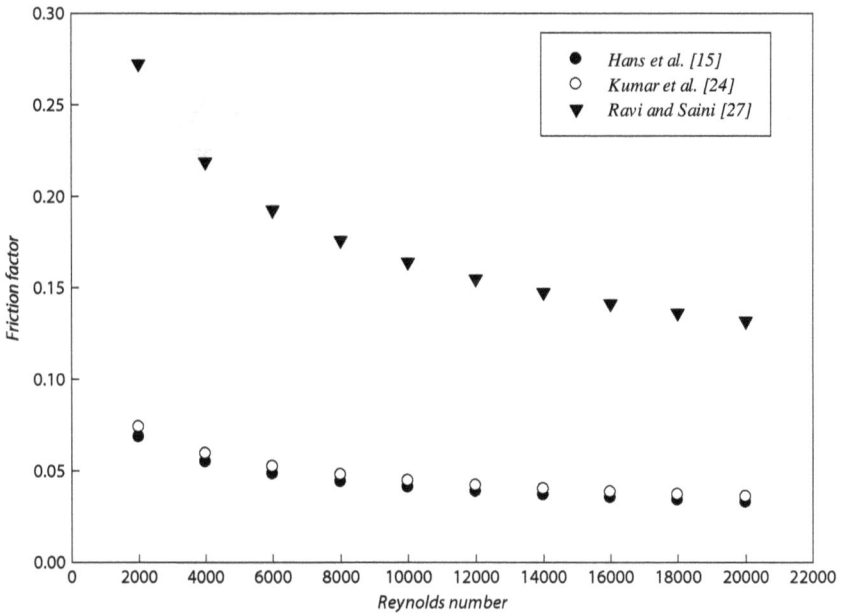

Fig. 23: Variation of Friction Factor with Reynolds Number for Different Multi V-rib Roughness Geometries in SAH

geometries as seen from Fig. 24, the multi gap V-rib combined with staggered element geometry of Deo et al. [26] has best thermo-hydraulic performance. Similarly among multi V-rib geometries, as seen from Fig. 25 the multi V-shaped roughness

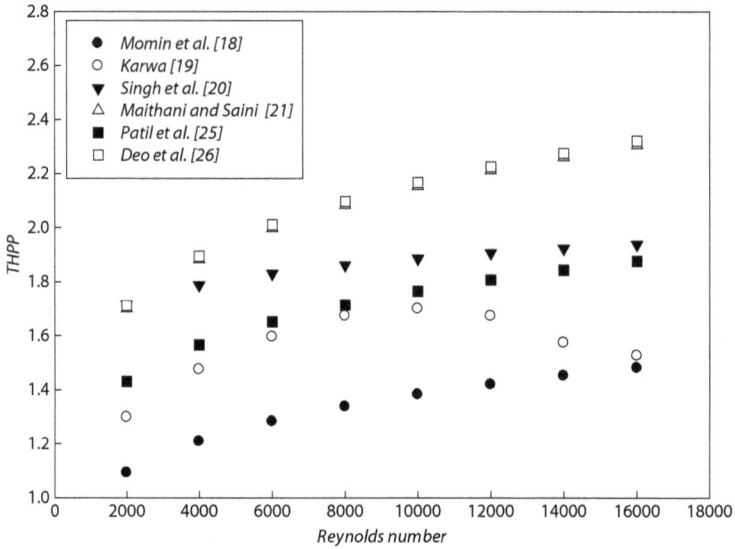

Fig. 24: Variation of Thermo-Hydraulic Performance Parameter With Reynolds Number for Different V-rib Roughness Geometries in SAH

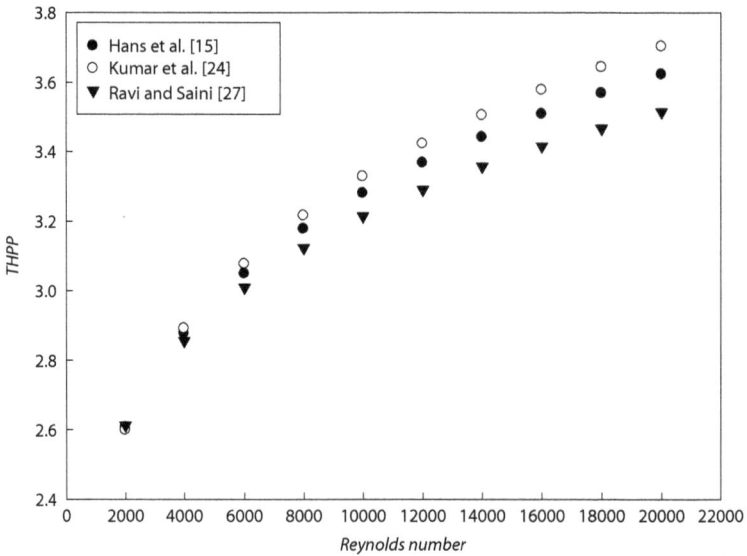

Fig. 25: Variation of Thermo-Hydraulic Performance Parameter With Reynolds Number for Different Multi V-rib Roughness Geometries in SAH

with a gap of Kumar et al. [24] has the best thermo-hydraulic performance

Further, among V-rib geometries, Jain and Lanjewar [28] and Patel and Lanjewar [29] both have reported experimentally better performance of their respective geometries as compared to multigap V-rib combined with staggered element geometry of Deo et al. [26]. Thus in V-rib geometry V-rib with symmetrical gap and staggered element geometry of Jain and Lanjewar [28] and gap in V-rib with symmetrical gap and staggered element geometry of Patel and Lanjewar [29] are better V-rib geometries till date.

7.0 Conclusions

A brief review of V-rib roughness geometries used in SAH has been presented. V-rib roughness pattern in various shapes have been investigated by many researchers and substantial enhancement in heat transfer have been reported. Few researchers have used CFD tool to study flow in SAH with V-rib artificial roughness and its effect on thermal performance. Based on a brief literature review following conclusions can be drawn.

(1) Using artificial roughness enhances the thermal performance of SAH. Thermal performance is improved due to the breaking of laminar sublayer.
(2) Roughness parameters such as relative roughness height, relative roughness pitch, angle of attack, relative gap width, relative gap position, relative roughness width, relative staggered element position, and relative staggered element size significantly affect flow pattern in SAH.
(3) Gap in V-rib significantly improves performance as compared to continuous V-rib due to flow acceleration in gap. Providing a staggered element further enhances performance due to scattering and double reattachment of flow. Combination of gaps, gap width, and staggered element significantly improves heat transfer but more specifically providing gap is more effective than others as

not only it accelerates flow through gap but also prevents redevelopment of boundary layer on rib element.

(4) As future scope for further development and in-depth study, CFD tool has to be used in conjunction with experimental methods as CFD provides fast and in-depth analysis of fluid flow that can be employed to prepare suitable roughness geometry.

(5) Based on correlations available in literature for V-rib geometries, the multigap V-rib combined with staggered element geometry has best thermo-hydraulic performance and among multi V-rib geometries multi V-shaped roughness with gap geometry has best thermo-hydraulic performance

As per recent experimental results reported in literature in V-rib geometry, the V-rib with symmetrical gap and staggered element geometry and gap in V-rib with symmetrical gap and staggered element geometry are better V-rib geometries till date.

Nomenclature

d	Gap position, m
D or Dh	Hydraulic diameter of duct, m
e	Rib height, m
g	Gap width, m
g'	Gap width of additional gap, m
G	Mass flow rate per unit area of plate, kg/sm2
H	Duct height, m
p	Pitch, m
p'	Staggered rib position, m
r	Staggered rib size, m
s	Length of main segment of rib, m
w	Roughness width, m
W	Duct width, m

Dimensionless parameters

B/S	Relative roughness length of discrete rib elements
d/w	Relative gap position
e/D or e/Dh	Relative roughness height
e+	Roughness Reynolds number
f	Friction factor
g/e	Relative gap width
Gd/Lv	Relative gap distance
Nu	Nusselt number
Ng	Number of gaps
p/e	Relative roughness pitch
p'/p	Relative staggered rib position
r/e	Relative staggered rib size
Re	Reynolds number
W/w	Relative roughness width
W/H	Duct aspect ratio

Greek symbols

α	Flow angle of attack, degree
δ	Thickness of laminar sub layer, m

Abbreviations

cw	Clock wise
ccw	Counter clock wise

References

(1) Varun RPS, Singal SK. A review on roughness geometry used in solar air heaters. *Solar Energy*. 2007;81(11):1340–1350.

(2) Duffie JA, Beckman WA. *Solar Engineering of Thermal Processes*. John Wiley & Sons: Hoboken; 2013.

(3) Singh S, Chander S, Saini JS. Thermo-hydraulic performance due to relative roughness pitch in V-down rib with gap in solar air

heater duct—comparison with similar rib roughness geometries. *Renew Sustain Energy Rev.* 2015;43:1159–1166.

(4) Maithani R, Saini JS. Heat transfer and fluid flow behaviour of a rectangular duct roughened with V-ribs with symmetrical gaps. *Int J Ambient Energy.* 2017;38(4):347–355.

(5) Kalogirou SA. *Solar Energy Engineering: Processes and Systems.* Amsterdam, the Netherlands: Academic Press; 2013.

(6) Jain SK, Agrawal GD, Misra RA. Detailed review on various V-shaped ribs roughened solar air heater. *Heat Mass Transf.* 2019;55(12):3369–3412.

(7) Prasad BN, Saini JS. Effect of artificial roughness on heat transfer and friction factor in a solar air heater. *Solar Energy.* 1988;41(6):555–560.

(8) Verma SK, Prasad BN. Investigation for the optimal thermohydraulic performance of artificially roughened solar air heaters. *Renew Energy.* 2000;20(1):19–36.

(9) Han JC, Glicksman LR, Rohsenow WM. An investigation of heat transfer and friction for rib-roughened surfaces. *Int J Heat Mass Transf.* 1987;21(8):1143–1156.

(10) Taslim ME, Li T, Kercher DM. Experimental heat transfer and friction in channels roughened with angled, V-shaped, and discrete ribs on two opposite walls. *J Turbomach.* 1996;118(1):20–28.

(11) Karwa R, Solanki SC, Saini JS. Heat transfer coefficient and friction factor correlations for the transitional flow regime in rib-roughened rectangular ducts. *Int J Heat Mass Transf.* 1999;42(9):1597–1615.

(12) Han JC, Park JS. Developing heat transfer in rectangular channels with rib turbulators. *Int J Heat Mass Transf.* 1988;31(1):183–195.

(13) Aharwal KR, Gandhi BK, Saini JS. Experimental investigation on heat-transfer enhancement due to a gap in an inclined continuous rib arrangement in a rectangular duct of solar air heater. *Renew Energy.* 2008;33(4):585–596.

(14) Singh S, Chander S, Saini JS. Heat transfer and friction factor of discrete V-down rib roughened solar air heater ducts. *J Renew Sustain Energy.* 2011;3(1):013108.

(15) Hans VS, Saini RP, Saini JS. Heat transfer and friction factor correlations for a solar air heater duct roughened artificially with multiple v-ribs. *Solar Energy.* 2010;84(6):898–911.

(16) Patil AK, Saini JS, Kumar K. Heat transfer and friction characteristics of solar air heater duct roughened by broken V-shape ribs combined with staggered rib piece. *J Renew Sustain Energy*. 2012;4(1):013115.

(17) Patel SS, Lanjewar A. Experimental analysis for augmentation of heat transfer in multiple discrete V-patterns combined with staggered ribs solar air heater. *Renew Energy Focus*. 2018;25(6):31–39.

(18) Momin AM, Saini JS, Solanki SC. Heat transfer and friction in solar air heater duct with V-shaped rib roughness on absorber plate. *Int J Heat Mass Transf*. 2002;45(16):3383–3396.

(19) Karwa R. Experimental studies of augmented heat transfer and friction in asymmetrically heated rectangular ducts with ribs on the heated wall in transverse, inclined, v-continuous and v-discrete pattern. *Int Commun Heat Mass Transf*. 2003;30(2):241–250.

(20) Singh S, Chander S, Saini JS. Heat transfer and friction factor correlations of solar air heater ducts artificially roughened with discrete V-down ribs. *Energy*. 2011;36(8):5053–5064.

(21) Maithani R, Saini JS. Heat transfer and friction factor correlations for a solar air heater duct roughened artificially with V-ribs with symmetrical gaps. *Exp Therm Fluid Sci*. 2016;70:220–227.

(22) Karwa R, Chitoshiya G. Performance study of solar air heater having v-down discrete ribs on absorber plate. *Energy*. 2013;55:939–955.

(23) Karwa R, Chauhan K. Performance evaluation of solar air heaters having v-down discrete rib roughness on the absorber plate. *Energy*. 2010;35(1):398–409.

(24) Kumar A, Saini RP, Saini JS. Development of correlations for Nusselt number and friction factor for solar air heater with roughened duct having multi v-shaped with gap rib as artificial roughness. *Renew Energy*. 2013;58:151–163.

(25) Patil AK, Saini JS, Kumar K. Nusselt number and friction factor correlations for solar air heater duct with broken V-down ribs combined with staggered rib roughness. *J Renew Sustain Energy*. 2012;4(3):033122.

(26) Deo NS, Chander S, Saini JS. Performance analysis of solar air heater duct roughened with multigap V-down ribs combined with staggered ribs. *Renew Energy*. 2016;91:484–500.

(27) Ravi RK, Saini RP. Nusselt number and friction factor correlations for forced convective type counter flow solar air heater having discrete multi V shaped and staggered rib roughness on both sides of the absorber plate. *Appl Therm Eng*. 2048;129:735–746.

(28) Jain PK, Lanjewar A. Overview of V-RIB geometries in solar air heater and performance evaluation of a new V-RIB geometry. *Renew Energy*. 2019;133:77–90.

(29) Patel SS, Lanjewar A. Experimental and numerical investigation of solar air heater with novel V-rib geometry. *J Energy Storage*. 2019;21:750–764.

(30) Rana J, Silori A, Maithani R, Chamoli S. CFD analysis of a V-rib with gap roughened solar air heater. *Therm Sci*. 2018;22(2):963–972.

(31) Jin D, Zhang M, Wang P, Xu S. Numerical investigation of heat transfer and fluid flow in a solar air heater duct with multi V-shaped ribs on the absorber plate. *Energy*. 2015;89:178–190.

(32) Kumar A, Kim MH. CFD analysis on the thermal hydraulic performance of an SAH duct with multi V-shape roughened ribs. *Energies*. 2016;9(6):415–438.

(33) Sharma AK, Thakur NS. CFD based fluid flow and heat transfer analysis of a v shaped roughened surface solar air heater. *Int J Eng Sci Technol*. 2012;4(5):2115–2121.

(34) Karanth KV, Manjunath MS, Sharma NY. Three dimensional CFD analysis of solar air heater for enhancement of thermal performance using arc shaped wire turbulators. ASME International Mechanical Engineering Congress and Exposition, IMECE, 2013;San Diego, CA.

(35) Dongxu J, Shenglin Q, Jianguo Z, Xu S. Numerical investigation of heat transfer enhancement in a solar air heater roughened by multiple V-shaped ribs. *Renew Energy*. 2019;134:78–88.

(36) Lewis MJ. Optimizing the thermohydraulic performance of rough surfaces. *Int J Heat Mass Transf*. 1975;18(11):1243–1248.

CHAPTER 5

Retrofitting Aging Tall and Supertall Buildings: A Continuing Challenge

Kheir Al-Kodmany

Department of Urban Planning and Policy, University of Illinois at Chicago, Chicago, USA

Abstract

Green retrofitting aging buildings is one of the cornerstones of sustainable development. This chapter investigates innovative technologies, trends, and practices of retrofitting tall and supertall buildings. It enlightens about practical design approaches, clarifies misconceptions, and offers useful directions. The chapter also reviews significant green retrofit projects of tall and supertall buildings and highlights the continuing challenges that building owners, architects, and engineers face in the path of green retrofits.[a]

Keywords: Building performance, energy consumption, environmental impact, remaining obstacles

[a] The author thanks the reviewers for useful feedback. Researchers faced serious difficulties to obtain data and updates on the performance of the retrofitted buildings. Parts of this chapter appeared in an Open Access Journal published by the author, with permission to reuse under the terms and conditions of the Creative Commons Attribution license (http://creativecommons.org/licenses/by/4.0/).

1.0   Introduction

A building is analogous to a living organism. It ages, gets sick and dies. Globally, many buildings have outlived their useful lives and have become inefficient [1]. It is estimated that older buildings in the United States consume 41% of the nation's total energy use, and in major urban centers, this figure surpasses 70% [2]. However, building owners are reluctant to replace them with new ones. They particularly hesitant to demolish tall and supertall buildings due to lack of expertise, the complexity of the process, and involved costs and risks. For example, the demolition of the 34-m-tall (110-foot-tall), 83-year-old Royal Canberra Hospital in Australia in 1997 resulted in injuring nine people and killing a child [3]. We may then contemplate about the costs and risks of demolishing taller buildings such as the 381-m-tall (1250-foot-tall), 89-year-old Empire State Building (ESB) in New York City.

Consequently, despite the growing stock of aging tall buildings,[b] only a small proportion has been demolished. The only building with a height greater than 200 m (656 ft) that was demolished is the 216-m-tall (709-foot-tall), 52-story JPMorgan Chase Tower in New York City. It was constructed in 1970 and was demolished recently to make room for a brand new 70-story tower.[c] The second tallest demolished building is the 178-m-tall (583-foot-tall), 41-story Singer Building in New York City in 1968. Constructed in 1908, it was replaced by the 227-m-tall (745-foot-tall), 54-story One Liberty Plaza in 1972.

Therefore, instead of going through a demolishing process, the industry is diligently working on finding cost-effective ways to extend the life span of tall buildings. With the advent of the green movement, new opportunities have arisen. Technological and scientific advancements that provide greater efficiencies and performances in conjunction with attractive financial incentives

---

[b] In this research, "tall buildings" refer to buildings with a height of 50 meters (164 feet) or greater and "supertall buildings" refer to buildings with a height of 300 meters (984 feet) or greater.

[c] The demolition of the World Trade Center does not count because it was a result of terrorist attacks on Sept. 11, 2001.

have encouraged building owners to green retrofit old buildings. The USGBC (US Green Building Council) explains that the 2009 market for major green renovations in the United States was $2.1 billion. It grew to over $6 billion by 2013 and to over $9 billion in 2019 [2]. In the United States, about half of commercial buildings were constructed before the advent of climate change, global warming, and sustainability,[d] and it is estimated that today over 50% of all construction projects engage retrofitting older buildings [2].

1.1   *Goals and Objectives*

Despite the rapid growth of the green retrofit industry, there is a continuing knowledge gap about this topic. Upon searching for information, we face difficulties in finding academic articles, for example. The Internet search results in finding a plethora of scattered pieces of media news, industry reports, and personal blogs that give a shattered understanding. There is considerable confusion about the virtues of green retrofitting older buildings, particularly tall buildings. There is a need to know how to decide when a building is vulnerable and when it is salvageable. Are retrofitting projects a mixed-bag endeavor? For example, in the case of the recently demolished JPMorgan Chase Tower in New York City, the building owner explained that despite a robust green retrofit that took place in 2012, in recent times, the building became insufficient to accommodate its growing needs. This review chapter attempts to tackle these dilemmas. It aims to shed light on the subject matter by carefully stitching and summarizing diverse resources.

The chapter consists of two parts. The first part answers basic questions such as "what is a green retrofit?", "what are the driving factors of a green retrofit?", "who is demanding a green retrofit?," and "what are the common features of a green retrofit?" The second part discusses practical green retrofit tall building projects that come from various parts of the world. The case studies address buildings of different uses and heights, including tall and supertall buildings that were green retrofitted in the past

---

[d] The approximate date for climate change and global warming starts in the 1980s.

two decades. Also, their ages vary from a few decades to almost a century. The intent is to make the case studies capture a diverse set of projects that help to identify the overlapping and varying retrofit activities. Finally, the chapter summarizes key findings and highlights continuing challenges.

1.2 What is a "Green Retrofit"?

A "green retrofit" could mean different things. A review of architecture literature reveals that architects and developers have been using the phrase "green retrofit" to refer to a wide-range of projects that involve for example installing a building automation system; replacing the insulation system; upgrading the heating, ventilation, and air conditioning (HVAC) system; strengthening the structural stability of a building; integrating light-emitting diode (LED); adding solar panels or wind turbines; incorporating a green roof or a green wall, and the like [1,2]. Among the prevailing practical views, however, is the one provided by the USGBC. It explains that a green retrofit is "any kind of upgrade at an existing building (EBs) that is wholly or partially occupied to improve energy and environmental performance, reduce water use and improve the comfort and quality of the space in terms of natural light, air quality, and noise—all done in a way that it is financially beneficial to the owner" [2].

Researchers see that a "green retrofit" serves sustainability's 3Rs, i.e., reduce, reuse, and recycle. By choosing to renovate over to demolish, we save on the building materials needed to construct a new building [2–4]. Further, a "green" retrofit makes the building more efficient on energy consumption, hence, reduces monetary costs and environmental damage. A green retrofit makes users enjoy more comfortable living and working spaces, amenities, and services. In short, a green retrofit could support sustainability's three spheres: the social, economic, and environmental.

From a practical point of view, it is useful to differentiate between two types of green retrofit: minor and major. A minor retrofit involves one type of work, e.g., mounting solar panels on a roof, upgrading the HVAC system, or placing a bike rack outside the building. In contrast, a major retrofit (or "deep retrofit") involves working on multiple systems related, e.g., to energy and utility efficiencies, renewable energy, structural stability, air quality, safety, and security. As such, undertaking a deep retrofit could be complex and may engage both interior and exterior spaces, making it technically challenging, costly, and a multiyear project. One of the possible choices is to conduct the retrofit in stages, starting with most needed and affordable to more expensive. For example, a retrofit project could start by upgrading the lighting system, later, replacing the HVAC system, and next, replacing windows, and so on.

2.0   Green Retrofit Drivers

Several factors stimulate green retrofitting buildings. Among these are boosting economic gain, improving environmental performance, taking advantage of governmental policies and financial incentives, joining the "new trend," and ensuring structural stability (Fig. 1).

Economic	Environmental	Financial	Fashion	Structural
• Reduce energy and water consumption • Increase the life span of the building • Decrease capital improvement program costs • Increase efficiency of interior spaces • Meet new functional needs • Decrease maintenance costs • Increase rent, occupancy rate, and tenants' retention • Obtain additional income from new facilities and amenities • Increase profit by improving employees' productivity • Save on building materials if the alternative was to construct a new building • Engender positive externality to neighborhood	• Improve health and living conditions, e.g., air quality, lighting level, thermal performance, acoustic, non-toxic materials, etc. • Reduce water consumption • Meet ADA requirements • Decrease GHG emission resulting from building operation • Save on embodied energy and GHG emission if the alternative was to replace the existing building with a new one • Use of recyclable materials	• Stimulus packages • Green programs • Sustainability initiatives • Rebates • Tax deductions, breaks, and credits • Attractive bank loans with low-rate interest • Government loans with long-term repay • Government grants	• Peer pressure • Market demand • Government programs • "State-of-the-art"	• Improve safety • Increase longevity • Avoid demolition

Fig. 1: Green Retrofit Projects: Key Drivers

2.1   *Economic Gain*

Simply, as buildings age, their performance and efficiencies decline, and their maintenance costs increase. Economic and energy crises also force buildings' owners to search for ways to reduce energy costs and use. Fortunately, new technologies motivate replacement for offering more efficient and less expensive operational costs. For example, many of the new lighting systems (e.g., LED) are more cost-effective than older ones. Also, newer systems may feature greater longevity. According to the USGBC, the average Leadership in Energy and Environmental Design (LEED)-certified building uses "32% less electricity and saves 350 metric tons of CO_2 emissions annually" [2]. Further, a substantial saving potential may be harnessed via the improved workforce productivity engendered by the healthier retrofitted spaces. Research explains that if an organization manages to boost employees' productivity by 10%, the resulting saving is likely to equal the cost of renting the space [5]. Further, green retrofitting could engender a positive externality. For example, the economic benefits of retrofitting downtown buildings of major cities—in many cases tall buildings—could permeate an entire downtown [6].

2.2   *Environmental Impact*

A massive driver of global heating is our buildings. Construction of new buildings and operations of existing ones are responsible for 39% of global energy-related emissions [2]. Surely, constructing and operating tall buildings consume enormous energy and generate immense carbon dioxide. Green retrofitting buildings offers clear environmental benefits, including decreasing energy consumption and reducing greenhouse gas (GHG) emission. For example, in New York City, GHG emission from its nearly one million buildings (where more than 5,500 of them are classified as tall buildings), constitutes almost 80% of the city's carbon footprint. In the case of the City of Portland, Oregon, research shows that if the city were to retrofit buildings that otherwise likely to be demolished over the next decade, the city will then reduce carbon dioxide emission by 231,000 metric tons, approximately 30% of the city's total carbon dioxide reduction targets [6,7]. Further, despite the potential benefits of new buildings (e.g., energy efficiency), it takes long years to offset the negative environmental impact resulting from the construction process [7]. As such, climate scientists view building retrofit as a good strategy for immediate response to climate change. In other words, building reuse can avoid unnecessary excessive carbon dioxide emissions and help cities achieve their short-term carbon reduction goals [7].

2.3   Governmental Policies and Financial Incentives

More governments see the economic and environmental values of retrofitting buildings [8]. For example, the U.S. government has been providing financial incentives to encourage green retrofit projects through offering tax breaks, credits, and grants. One of the earliest and ambitious green programs that encouraged retrofitting is the Energy Efficiency Building Retrofit Program launched in 2007 by a former U.S. President William Clinton. In 2011, President Obama's "Better Buildings Initiative" proposed new efforts to improve energy efficiency in commercial buildings across the U.S. President Obama reasoned that improving the energy efficiency of buildings can create jobs, save money, reduce the country's dependence on foreign oil, and make the air cleaner. He supported the program throughout his administration, ending in January 2017. Funding opportunities for green retrofitting buildings continue to be offered by several U.S. agencies and organizations, e.g., the Environmental Protection Agency (EPA), and the USGBC [8–10].

2.4   A New Trend

Peer pressure or "keeping up with the Joneses" is another playing factor. For example, the green retrofit of the ESB has stimulated other important buildings such as the Willis Tower (formerly Sears Tower), Chicago's tallest building. Interestingly, it is becoming fashionable to locate in "green" buildings. For example, after retrofitting the ESB, the Federal Deposit Insurance Corporation (FDIC) and Skanska USA, the U.S. division of Swedish construction firm Skanska AB committed to major leases. Indeed, high-profile tenants increasingly desire to locate in a LEED-certified buildings [9,10]. Realtors anticipate that this phenomenon will increase as people are becoming more conscious about the economic and environmental benefits of green design. Four types of tenants are at the forefront in demanding green workplaces [10–12]:

- The Fortune 500 multinational corporations
- The "gazelles," i.e., the companies that want to recruit the "creative class" who sees the value of "green" buildings
- Government tenants who are pushing the demand because their own policies require such facilities
- The public sector at large who is increasingly boosting efforts to pass new sustainable development legislation and retrofit public buildings with energy-efficient features [10].

2.5 *Structural Stability*

Many tall buildings were constructed before modern seismic codes and wind resistance standards. With the advent of climate change, buildings have to be prepared to face new powerful threats and hazards. In response, today, some cities are mandating retrofitting tall buildings to improve their structural performance. For example, recently, the City of San Francisco, California, is working on upgrading their seismic standards after releasing a list of 150 buildings that are entitled for structural retrofit. This act was triggered by the 58-story Millennium Tower in the city that suffered from serious sinking and tilting.[e] Likewise, the City of Tokyo embarked on a plan to retrofit skyscrapers in preparation for anticipated major earthquakes. Based on this initiative, several buildings (e.g., Shinjuku Mitsui Building) were retrofitted with damping systems because they were constructed before these

---

[e] The incidence did not result only in vacating the building but also temporarily shut down the nearby $2.26 billion Transbay bus station, completed recently.

systems were common[f] [11–14]. In addition to occupants' safety, the "green" aspects of a structural retrofit refer to the argument that the total costs and carbon dioxide emission resulting from demolishing an older building and constructing a new one would be far greater than that resulting from retrofitting it [15–17].

3.0 Common Practices of Green Retrofit

While there are numerous ways to retrofit a building, the following outlines some of the most prevailing ones [18–23].

3.1 Lighting Systems

The most prevailing retrofit type is the lighting systems retrofit. Upgrading lighting fixtures can result in an increase in the lighting level while decreasing energy consumption up to 70%, yielding significant cost savings [3]. Besides, a lighting system retrofit is one of the easiest to conduct because it entails little or no interruption to the building's daily activities. Here are some options for a lighting retrofit project [3]:

- Change the old fluorescent lighting fixtures into Energy Star benchmarked fixtures, e.g., T5 or T8 high bay fixtures
- Add a timer or occupancy sensor on the fixtures that are only used occasionally, allowing the lights to be turned off automatically when it is not in use
- Add a dimmer or photo-sensor for the fixtures so that when natural light is available, photo-sensors will adjust the brightness of the fixtures to reduce unnecessary lighting [3].

---

[f] Mass dampers are gigantic pendulum-like counterweights that mitigate the vibration impact by pulling a building's mass in the opposite direction of the prevailing forces.

3.2 HVAC Systems

Newer HVAC systems can result in improving tenants' comfort
and minimizing environmental impact. Best practice recommends
cleaning air filters, ventilators, boiler tubes, and sealing all the
HVAC equipment so that heat transfer occurs only when and
where it is desired [3].

3.3 Water Systems

With dramatic urban population increase, potable water is
increasingly a precious commodity. As such, water conservation is
becoming a priority. The following specific strategies are helpful
to conserve water [3]:

- Upgrade faucets, toilets, and showerheads fixtures that
  were made before 1992. Since 1992, the U.S. federal
  legislation requires that toilet, faucet, and showerheads to
  utilize at most 1.6 gallons per flush (gpf), 2.5 gallons per
  minute (gpm), and 2.5 gpm, respectively
- Upgrade to waterless urinal for it uses sealant liquid that
  has higher buoyancy than urine
- Consider adding aerators and occupancy sensors on
  lavatory faucets for they reduce the rate of water flowing
  through the faucets by mixing water with air while
  maintaining the pressure of the water
- Consider reducing water use by recycling it
- Rainwater can also be captured for irrigation or even to
  flush toilets and other uses in the building

3.4 Building Automation System

A building automation system manages all the operation systems,
including the HVAC system, and the lighting system, as well as
appliances. It helps to reduce utility costs and maintenance while
improving user comfort. For example, the system can maintain
the temperature, air quality, and lighting inside the building based

on a given range and preset schedules. In addition, the system can accurately monitor and record energy usage. It also helps to identify and locate a problem if unusual energy use or system failure occurs. Further, the building automation system can be commanded remotely online and allows immediate intervention in emergency cases.

3.5 Insulation System

Through robust insulation, a building can significantly reduce the need for heating and cooling, leading to significant reduction in energy consumption and carbon dioxide emission. For example, well-insulated windows can help in reducing heat transfer between 40% and 70% of a building, thereby reducing energy costs ([3], p.11). In the case of the ESB (see the following section), upgrading window insulation helped to save $410,000 annually [13]. Building owners may consider the following strategies to enhance window insulation:

- Replace existing windows with low U-factor[g] windows and add weather-stripping[h] on windows to prevent air leakage
- Replace the single-pane windows with double-pane windows
- Apply low emissivity coating on the windows to further lower heat transfer between inside and outside [3]
- Pick window frames that have a low U-factor.

An insulation upgrade for walls is uncommon because it is costly. An affordable way to make walls less absorbent of heat is by painting them lighter colors for they reflect light effectively. Further, replacing a single door entrance with a double one (with weather-stripping) is often an affordable and effective insulation measure [3].

---

[g] U-Factor measures the rate of heat transfer and ranges from 0.25 to 1.25 where the lower value indicates better window insulation.
[h] Weather-stripping can also be applied to doors to improve insulation.

3.6   On-Site Energy Generation

The fossil fuel that we heavily rely on today to generate energy is a finite source, and it is crucial that we seek alternative sources. Solar photovoltaic (PV) and solar thermal (the latter is more affordable) are increasingly popular. Wind turbines are also becoming available and affordable; which can be incorporated in retrofits. For example, the Willis tower retrofit project plans to incorporate wind turbines to harness wind energy. Geothermal energy[i] could be cheaper than the solar one but are available in limited geographic areas. Ultimately, the right choice of energy renewable adoption depends on the location and climatic conditions of solar intensity, wind power, humidity, cloudiness, and particles in the air [3].

3.7   Technological Innovations

Technological and scientific advancements provide planners and architects new ways to improve the environmental performance and energy efficiencies of buildings. For example, lately, 3M (a global innovation company, located in Minneapolis, Minnesota, United States) has introduced 3M solar film that can be placed on windows to generate energy and reduce the energy needed for cooling buildings by absorbing more than 90% of infrared light. This thin solar film has started to be available in the market at affordable prices. However, the downside of this film is that it provides only 3%–8% of efficiency during peak intensity, measuring 20% of what conventional solar PV can generate [12,13]. Another example that highlights technological advancement is the Eco-Skin, a lightweight, transparent textile that can improve a building's insulation and generate electricity by wind or sunlight [15].

---

[i] Geothermal energy is heat derived by water or stream from the earth sub-surface. Depending on its characteristics, geothermal energy can be used for heating and cooling purposes or be harnessed to generate clean electricity.

4.0   Retrofitting Tall Buildings: Case Studies

4.1   ESB, "The Green Empire," New York City, United States

The ESB is a 102-story, 381 m (1,250 ft) skyscraper located in Midtown Manhattan, New York City. The building's construction was completed in 1931, and it remained the tallest building in the United States for almost four decades, until the topping out of the World Trade Center's North Tower in late 1970. With its beautiful Art Deco style, elegant profile, and distinctive height and history, the building enjoys an important place in the American culture as a symbol of the power of the State of New York. It was named as one of the Seven Wonders of the Modern World by the American Society of Civil Engineers. Its 260,128 m^2 (2,800,000 ft^2) of leasable office space attracts a wide-range of tenants and businesses, drawn by the building's prestige, unmatched skyline views, and convenient location at the center of Manhattan's mass-transit system. Its observatory on the 86th floor attracts between 3.5 and 4 million visitors yearly. The ESB that for long represented a symbol of the power of the New York City has become recently also a symbol of "green power efficiencies," or what was nicknamed "the Green Empire" [15,18].

The green retrofit project began in 2009 as part of the Clinton Global Initiative, and in 2010, the then 80-year-old building underwent an over half-billion-dollar retrofit, with the goal to transform the building into a more energy-efficient and eco-friendly structure. The building upgrade is the largest in the United States, leading to energy saving estimated by $4.4 million annually and cutting carbon dioxide emissions by 7,000 metric tons yearly [24–27]. As such, reducing the building's carbon footprint for 15 years is equivalent to removing 20,000 cars off the road. In terms of economic feasibility, the expected income stream enhancements include [28]

- Decreasing capital improvement program costs
- Reducing utility budget due to achieving higher efficiencies in energy and water usage
- Decreasing building operations budget due to lower maintenance and repair costs
- Increasing rent and occupancy rate due to providing higher-quality spaces of greater services and amenities
- Generating additional incomes from new facilities, amenities, and tenant services [28]

Making the ESB a green catalyst was an important goal of the project. It meant to help in changing building owners' perception of green retrofit as an expense, rather than an opportunity for economic gain and environmental improvement. In other words, the project is meant to highlight that a green retrofit makes an economic sense ([15], p. 56). A partnership of several organizations made the ESB's retrofit project possible, including the Clinton Climate Initiative (CCI); Jones Lang LaSalle; Rocky Mountain Institute (RMI); Johnson Controls Inc.; and ESB Operations.

The project team has pursued a systematic multiphase analytical examination. It conducted comprehensive analyses to determine which energy and sustainability strategies could be implemented and identified the associated costs, risks, and obstacles that might arise. Specifically, the project team examined the building's mechanical systems and equipment, computed tenants' energy consumption, and developed a baseline energy benchmark report and a system for gaging energy efficiency. Analyses suggested that the team should pursue a program that "would reduce energy use and greenhouse gas emissions by 38%, saving 105,000 metric tons of carbon dioxide over the next 15 years" ([28], p.8) (Fig. 2).

In addition to reducing energy and carbon dioxide emissions, the sustainability program planned to deliver an enhanced environment for tenants, including improved air quality resulting from tenant demand-controlled ventilation, better lighting conditions that coordinate ambient and task lighting; and improved thermal comfort resulting from better windows, new radiative barriers, and better HVAC system control. The upgrade engendered significant energy savings [29] (Figs. 3 and 4).

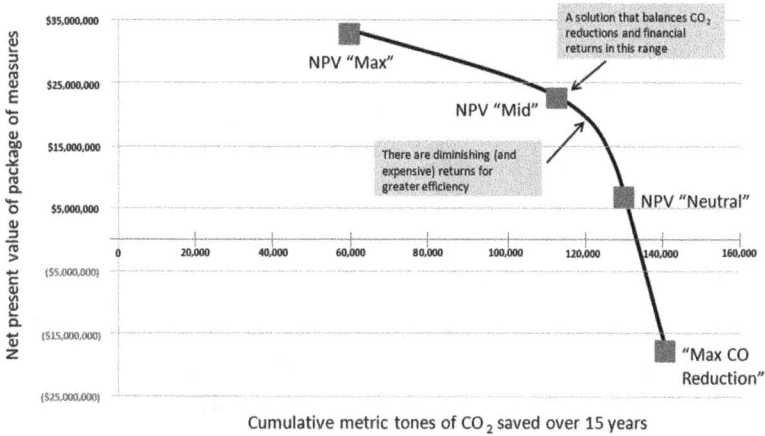

Fig. 2: Fifteen-Year Net Present Value (NPV) of Package *vs.* Cumulative CO_2 Savings. The Graph Demonstrates that the ESB can Achieve a High Level of CO_2 and Energy Reduction Cost-Effectively (Graph by Author; Redrawn From [28]).

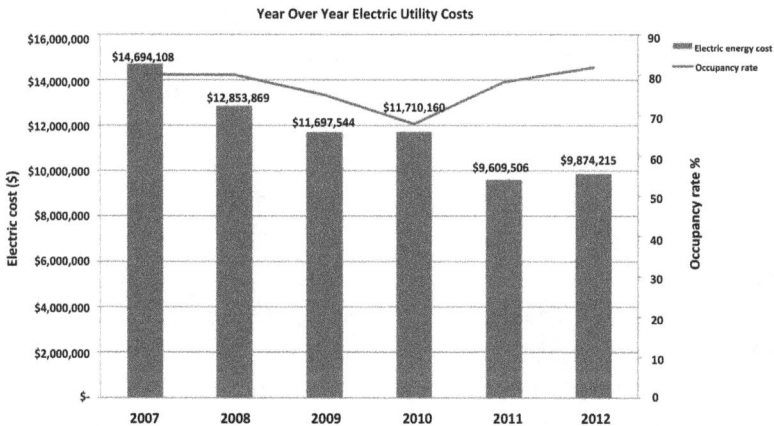

Fig. 3: ESB Performance Year 2. Reduction in ESB's 2007 Baseline Electric Utility Costs during Performance Period (Graph by the Author; Redrawn From [29]).

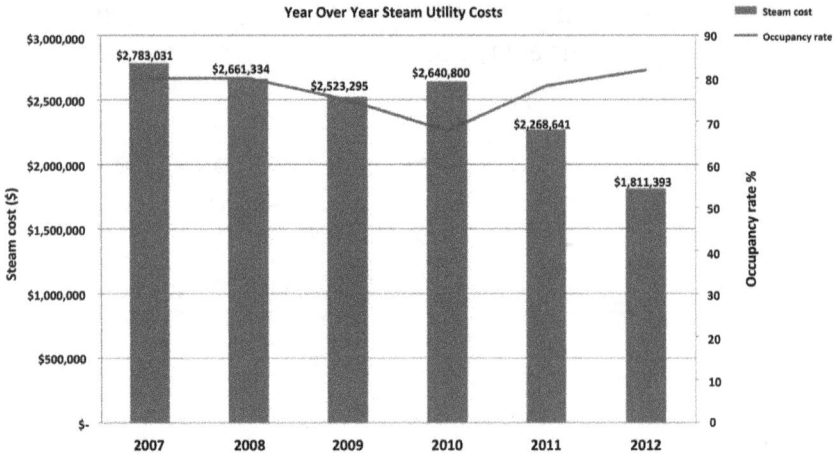

Fig. 4: ESB Performance Year 2. Reduction in ESB's 2007 Baseline Steam Utility Costs During Performance Period (Graph by the Author; Redrawn From [29]).

Among the most noteworthy retrofit aspects were windows and elevators. It is unprecedented that a skyscraper of this scale reuses, rather than replaces, about 96% of its window glass. This choice has saved the building's owners $2,300 per window and avoided the negative environmental impact of transporting new windows from the manufacturing plant and old ones to recycling factories. Also, it is remarkable that the window retrofit happened without disturbing tenants for windows were removed after office hours and reinstalled before most tenants returned to work the next morning.

Upgrading the elevator system was also remarkable because the new system saves energy in multiple ways. The employed regenerative system harvests the "waste energy" from braking, of which elevators do frequently. Conventional elevator machinery can lose more than 30% of its energy in the form of waste heat, and the new retrofit reduced the loss to only 5%. The second way of savings is a direct result of the first. In a conventional elevator system, waste heat gathers in the machine room, which then requires substantial air conditioning to prevent overheating. In contrast, the regenerative system does not have this problem because the heat has already been captured, harnessed, and channeled to the electrical system as "surplus" power [30,31]. Additional savings are achieved by having the new system's motor consumes zero energy when the elevator is not in use. Finally, the new elevator system is designed to accommodate high-efficiency LED lighting in the cabs. Collectively, the new elevator system results in significant energy savings that reduce demand on the city power grid.

Further, the building's owners signed an agreement with the Green Mountain Energy Company to purchase 100% of its power from renewable sources, resulting in a positive ecological impact. "By purchasing nearly 55 million kilowatt-hours of renewable energy each year, nearly 100 million pounds of carbon dioxide will be avoided annually" [32]. Most importantly, during the retrofit project, the ESB team engaged tenants in the planning stage and enlightened them about retrofit options to create energy-efficient office spaces that they like and enjoy. For example, one of the tenants designed an open office floor plan to maximize natural light and installed an under-floor ventilation system to bring in fresh air that allowed individual temperature controls for workspaces [32,33].

In addition to engaging tenants, the retrofit project embraced an integrated process that coalesced efforts of various project teams such as the capital and sustainability teams and facilitated a whole-building retrofit approach. For example, initially, the capital team has assigned a budget to replace the chiller plant. However, the sustainability team found out, based on a study it conducted, that chiller replacement could be avoided by retrofitting windows on-site and by just upgrading the existing chiller. Such a decision resulted in significant savings, for the chiller upgrade is far less expensive than replacing it [34].

The green retrofit of the ESB attained the LEED® EBOM (Existing Buildings: Operations and Maintenance) Gold certification. While the LEED-Gold certification applies to the entire structure as an EB, specific spaces within the building received LEED-Platinum certification, the highest designation under the USGBC standard for commercial interiors. The building also has earned a score of 90 (out of 100) from the Environmental Protection Agency's "Energy Star" program. Skanska has earned a LEED–Commercial Interiors Platinum rating for its 2,200 m² (24,000 ft²) offices on the 32nd floor [3,34]. In conclusion, the retrofit of the ESB offers a stimulating prototype for commercial office supertall buildings [34,35].

4.2 Willis Tower, Chicago, Illinois, United States

The 110-story, 442 m (1,451 ft) Willis Tower (formerly Sears Tower, 1973–2009), is a Chicago icon and the second tallest building in the Western Hemisphere after One World Trade Center in New York City. Completed in 1973, this supertall is a major tourist attraction in Chicago. Its observation deck attracts over one million visitors yearly [15,36]. Designed by the firm of Skidmore, Owings, and Merrill, the building enjoys an innovative structural system, developed by structural engineer Fazlur Rahman Khan, and a striking exterior of black aluminum and bronze-toned glass. Willis Tower has recently undergone a long-term green retrofit project by partnering with Smith and Gill architectural firm and Environmental Systems Design, a mechanical engineering firm. The retrofit plan projected "Willis Tower's base building energy use to be reduced by 80% (equal to 64M kWh annually) and its water use by 24 million gallons annually" [36]. The cost of the retrofit is estimated at $350 million.

The greening and modernization plan of the Willis Tower includes several features. The major energy saving for this renovation project is replacing the 1973 tower's 16,000-tinted single-pane windows with insulating glass units (IGUs), also known as insulated glass [37]. An IGU consists of two or more glass window panes separated by a vacuum or gas-filled space to reduce heat transfer. This retrofit measure is crucial to insulate the building from Chicago weather that fluctuates between extreme temperatures of cold winter and hot summer. Traditionally, insulated glass involves doubling or tripling the glass panes. However, for the Willis Tower, this option would be inappropriate for that will add substantial load on the curtain wall. Instead, the plan suggested providing the insulation through adding insulating thin-film that will have the insulating properties of triple-pane glazing while avoiding the heavyweight of triple glass. It is estimated that the reglazing retrofit will provide effective daylighting and save energy consumption required for cooling and heating the building [36].

Further, the Willis Tower retrofit project suggests upgrading the building's mechanical systems by supplying new gas boilers that utilize fuel cell technologies to generate electricity, and to heat, and cool the building with 90% efficiency. The project will also upgrade the building's 104 high-speed elevators and 15 escalators to reduce energy consumption. Further, it will install advanced lighting control systems that automatically dim the lights and adjust to an optimal level of brightness when daylight is detected, reducing energy consumption as well. The tower's retrofit will also involve upgrading the building's plumbing system and restrooms by installing low-flow water fixtures on toilets, urinals, and faucets; as well as by providing condensation recovery and irrigation systems. Drawing and renderings also show integrating green roofs as well as wind turbines and PV panels to harness wind and solar energy.

Similar to the ESB, the Willis Tower is an important national and international landmark supertall building that will likely inspire other buildings to conduct green retrofits despite the involved initial costs. Chicago Loop alone contains hundreds of tall buildings that are due to retrofit. Further, retrofit projects may enhance the economy by creating jobs; e.g., it is estimated that the retrofit project of the Willis Tower will create 3600 jobs [37].

4.3   Taipei 101, Taipei, Taiwan

Taiwan's Taipei 101 skyscraper—the world's tallest building from 2004 to 2010—has recently undergone a major green retrofit that made the tower to earn LEED Platinum certification—the highest level of achievement in the LEED system. A 3-year-long green retrofit has enabled this supertall building to achieve significant savings on electricity and water and to reduce and recycle waste. The retrofit has resulted in reducing annual utility costs by $700,000 a year and reducing the carbon dioxide emissions by nearly 3,000 tons per year; this is equivalent to removing 240 vehicles off the road [38,39]. The tower's retrofit involved upgrading its electrical and mechanical systems of HVAC, as well as enhancing its already efficient Siemens Apogee building management system so that it can carry out more accurate monitoring and analyses of energy consumption. Temperature and humidity sensors were installed on each floor so that they transmit information to the management system, which then decides on turning on and off mechanical and lighting equipment. Further, similar to the case of Bank of America Building, the building was equipped with "ice batteries" that make ice at night, when electricity is cheaper, and then melts during the day to cool the building in hot summers. In addition to upgrading these major systems, the retrofit also advanced other important features of the building including the irrigation system and food-waste recycling system. This is significant since the tower houses hundreds of restaurants. Collectively, these enhancements augmented the existing green features of the building, mainly low-E glass and greywater systems.

4.4   *Adobe World Headquarters, San Jose, California, United States*

Between 2000 and 2006, software maker Adobe engaged in a major retrofit of its headquarters, located in Downtown San Jose, California, United States. The over 92,903 m^2 (1,000,000 ft^2) complex comprises three commercial office high-rise buildings: West Tower, East Tower, and Almaden Tower, completed in 1996, 1998, and 2003, respectively. These towers rest atop 87,187 m^2 (938,473-ft^2) parking garage. In 2001, because of experiencing rolling blackouts and spikes in energy prices, the California government asked commercial users to reduce energy usage by 10%. In response, Adobe has embarked on a green retrofit project that earned the building Platinum LEED certification in 2006. The retrofit project resulted in reducing electricity use by 47%, gas use by 42%, and water use by 48%. Adobe's retrofit project cost was about $1.4 million but it received $389,000 in rebates and achieved a saving of $1.2 million a year, translating to a return on investment (ROI) of 121% [3,40].

The first step that Adobe took in the upgrade process was identifying utilities and equipment that were overusing energy. For example, the parking garage's exhaust fans were operating unnecessarily 24/7. The retrofit's preliminary study suggested that the fan running time could be reduced to only 3 hours during the morning commute and 3 hours during the evening commute—this will suffice to keep air quality above the required standards. Another important improvement involved providing a Web-Based Intelligent Building Interface System (IBIS) that monitors and controls the building's equipment. The IBIS comprises 30,000 monitoring points and is run by a single software program that would ease the tasks of monitoring building's subsystems by displaying electricity, water, gas, Uninterruptible Power Supply (UPS) systems, data centers, and standby generators status in real-time. The IBIS also allows adjusting the lighting and temperature of the numerous zones of the building individually, floor-by-floor, or the entire building remotely. Further, the IBIS helps in detecting and correcting problems ahead of time, resulting in significant annual savings [40].

The building's upgrade also involved implementing water conservation strategies. For example, the retrofit project improved restroom facilities by installing waterless urinals, automated flush valves, faucets, and soap and paper towel dispensers. The relandscaping scheme also emphasized using local and drought-tolerant plants to reduce water and maintenance needs and upgraded the subsurface drip irrigation system (which is more efficient than the spray irrigation system). Interestingly, Adobe already uses two satellite-based evapotranspiration (eT) controllers to regulate irrigation by communicating with local weather stations through wireless technology [41]. As such, the system optimally adjusts water flow according to local weather. Finally, Adobe has installed a web-based system to control pump-run time for the fountains.

The retrofit project also upgraded chillers by installing an adaptable frequency drive (AFd), which resulted in savings totaling approximately 300,000 kWh ($39,000) annually [3]. Adobe also employs effective composting and recycling of paper, cardboard, plastic, glass, cans, printer toner, batteries, kitchen grease, and the like, which result in diverting up to 95% of its solid waste from landfill. Adobe also employed green cleaning methods and materials that reduce health and environmental risks. These green materials are nontoxic, environmentally safe, VOC[j]-compliant, and biodegradable. Further, Adobe has encouraged using green transportation means. It installed locked bike cages and offered its employees incentives to use public transport. As a result, about 20% of the employees commute by public transport, compared to 4% in Silicon Valley [41].

4.5   Glastonbury House, London, United Kingdom

While there is an increasing number of commercial skyscrapers that embrace green retrofit, there are fewer residential buildings that take this path. As such, the retrofit project of the 23-story Glastonbury House, a residential building in London, United Kingdom, provides a good working model for other residential towers. The building has recently undergone a major 17 million dollars retrofit that resulted in 50% in energy saving and 40% in water reduction [3]. The retrofit project involved upgrading bathrooms (dual flush toilets and spray taps were installed to reduce waste of water), replacing light fixtures to increase lighting levels, and improving the heating systems' efficiencies. The project also involved installing renewable energy systems including solar PVand wind turbines to generate power, and solar water heaters to provide hot water. Further, eco-friendly materials were utilized and recycling was encouraged.

---

[j] Volatile Organic Compound

4.6   The Joseph Vance Building, Seattle, Washington, United States

After acquiring the 14-story historic Joseph Vance Building in downtown Seattle in 2006, the Rose Smart Growth Investment Fund I, L.P. has embarked on a series of retrofit projects intended to transform the building into "the leading green and historic class B building in the marketplace." The 12,400 m^2 (134,000 ft^2) terracotta Vance Building was constructed in 1929 and contains offices over ground-floor retails that had the potential to ignite economic activities and reenergize social and pedestrian life. The purchase was motivated by the company's mission to acquire assets in walkable, mass-transit accessible locations that are ripe for repurposing via green retrofits.

Key retrofit strategies included [42,43]:

- Replacing ducted heating system+;
- Recalibrating steam heating system
- Localizing thermostats
- Restoring windows to facilitate natural ventilation
- Adding automated sunshades to reduce heat gain
- Providing lighting retrofit with automated controls, and adding light shelves and CO_2 sensors
- Replacing the roof with a LEED-approved one
- Replacing water fixture
- Providing bike storage and shower facilities

The historic Vance Building already enjoys a great character and inherent environmentally sensitive features including high ceilings, operable windows, and floor plans designed to maximize natural light. The renovation work capitalized on these features by focusing on providing natural ventilation and daylighting. Modern technologies implementation was meant to complement these "vernacular" systems. The project team conceptualized a natural ventilation strategy to meet tenant thermal comfort needs in the summertime. The strategy was backed up with quantitative analysis, temperature monitoring, and targeted façade solar gain studies. For economic reasons, the project did not replace the existing steam heating system, which is connected to the Seattle Steam downtown grid, rather it upgraded it. Also, a replacement of the heating system would have involved substantial alteration to the existing structure.

The project also replaced the global thermal control with local controls to enable regulating the steam system at the individual radiator units. This measure will enhance user comfort and improve energy efficiency. Further, the 45-foot (14-m) building width facilitated natural lighting. This is also augmented by the light shelves on the south and west exposures, which reflect sunlight to light-colored ceilings. A lighting retrofit involved replacing inefficient fixtures with T8 and T5 fixtures [43]. Also, occupancy sensors were provided in all common areas and most tenant spaces. Further, the owner has mandated green cleaning and designed a green tenant improvement and operation manual to educate tenants on building green features and to inform about expected behaviors while using the building's utilities, spaces, and amenities.

The building's energy consumption prior to the retrofit was about 51k Btu/sf/yr, and after the renovation, it was dropped by 24%, to just 39k Btu/sf/yr [43]. The building now uses 58% less energy per square foot than the average for offices in the Unuted States. [43]. The project retrofit has improved the building's imageability; and consequently, the occupancy rate was increased from 68% to 96%, and the building has seen increased rents, tenant retention, and net operating income. The building's acquisition cost was $23.5 million ($176/sf); the improvement cost was $3.5 million ($26/sf), and the cost of the Tenant Improvements and Leasing Commissions was $2.26 million ($17/sf) [43] (Fig. 5).

The U.S. Green Building Council (USGBC) awarded the Vance Building a LEED for EEBs Gold certification. The building also achieved an Energy Star rating of 98 (out of 100). This placed the building in the top 2%of office buildings nationally. A Post Occupancy Evaluation (POE) has indicated that "77% of building occupants are satisfied with lighting levels. 85% of occupants indicated general satisfaction with the overall building and individual workspaces" ([43], p. 30). A study, conducted

Energy Use per square foot Comparison

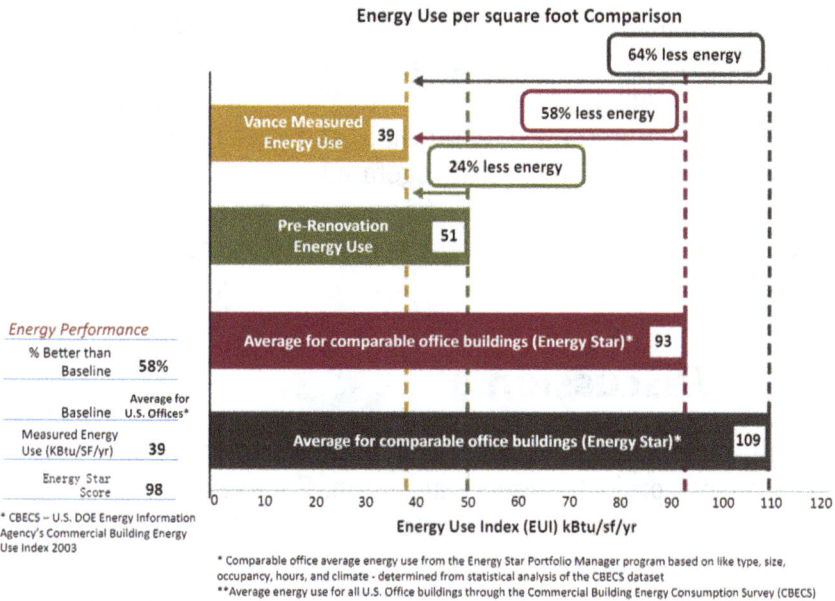

Fig. 5. Gauging Energy Performance of the Retrofitted Joseph Vance Building, Seattle, Washington, United Statrs (Diagram by the Author; Redrawn From [43]).

by the University of California, Berkeley Center for the Built Environment, also showed that tenants were mostly pleased with the new real-time energy monitoring system that helped them to conserve energy. They were also happy with the provided thermal comfort and acoustic.

4.7 Hanwha Headquarters, Seoul, Korea

Hanwha is one of the world's leading companies that produce environmentally-friendly products including PV panels. Recently, the company has felt that its 1980s-era headquarters building does not reflect its environmental values and image and consequently wanted to renovate its building. After soliciting for renovation proposals, Amsterdam-based UNStudio's proposal won the bid. The proposal involves replacing the building's façade of opaque paneling and dark single-pane glass windows with animated LED panels. The new façade also integrates a shading system. Where providing shade to the facade is desirable, the glazing is angled away from direct sunlight; while in other places the glazing is purposefully angled to receive direct sunlight [44]. In addition to façade replacement, the renovation project included remolding meeting rooms, auditorium, and executive areas, incorporating sky gardens and vegetated lobbies, and improving interior climate and air quality. As such, the renovation project also involved incorporating green walls and enhancing outdoor landscaping.

5.0 Discussion

Overall, the examined case studies in this chapter indicate that green retrofit projects deserve attention. For example, providing energy-efficient windows and installing efficient lighting systems could decrease demand on the city grid and reduce GHG emissions. Most skyscrapers enjoy large envelopes, which provide the potential for harnessing solar power. On a city-scale, this could offer quite a boost to the electrical grid. As mentioned, the Willis Tower in Chicago and Hanwha Headquarters in Seoul, South Korea, are embracing this approach as they move toward sustainability. Also, among the common retrofits is swapping buildings' old HVAC equipment for systems that generate ice during the night (when electricity rate is cheaper) to cool the building during hot daytime [45]. Table 1 summarizes the green retrofit features in the examined case studies.

While this chapter concerned mainly projects that tackled major retrofits (or "deep retrofits"), i.e., projects that involved several renovations work to improve the performance of the building; it is important to note that there are numerous green retrofit projects that deal with only a few renovations. Due to limited needs or restrained budgets, a building's owners may choose a simple retrofit project that yields substantial savings and productivity. For example, recently, the Sheraton Chicago Hotel and Towers installed motion-sensing thermostats in every room of the hotel's 1,214 rooms. It is a simple concept but it is resulting in a significant energy saving; it decreased the hotel's electric bills about $150,000 per year [14]. The way it works is that rather

Table 1: Major or "Deep" Green Retrofit of Tall Buildings: Summary of Project Examples (Source: developed by the author)

Building Name, Location, Date of Completion	Height (Architectural), Number of Floors Above Ground, and Gross Floor Area (GFA)	Green Retrofit Features Employed	Gained Benefits	Awards
ESB New York City, United States 1931	381 m/ 1,250 ft 102 stories 208,879 m²/ 2,248,355 ft²	Refurbished approximately 6,500 thermopane glass windows	Improved building's insulation and reduced energy consumption	LEED® EBOM (Existing Buildings: Operations and Maintenance) Gold certification to the entire structure as an existing building (EB) LEED-Platinum certification to specific spaces within the building
		Installed reflective barriers behind each radiator	Reduced energy consumption	
		Installed photo-sensors and dimmable ballasts so that when there is natural sunlight, artificial lighting can be turned on but dimmed to the desired level	Reduced energy consumption	

(Continued)

Table 1: Major or "Deep" Green Retrofit of Tall Buildings: Summary of Project Examples (Source: developed by the author) (*Continued*)

Building Name, Location, Date of Completion	Height (Architectural), Number of Floors Above Ground, and Gross Floor Area (GFA)	Green Retrofit Features Employed	Gained Benefits	Awards
		Replaced the tubes, valves, motors, etc. of the four chillers while retaining the shells	Improved system efficiency	
		Replaced over 300 existing air-handling units	Reduced energy consumption and improved air quality	
		Upgraded the EB control system to optimize HVAC operation	Reduced energy consumption	
		Installed a demand control ventilation system to ensure adequate ventilation by measuring the carbon dioxide concentration in the tenant space	Improved air quality	
		Installed individualized, web-based power usage systems for each tenant	Improved management capabilities	
		Installed the XeroFlor Green Roof System for four rooftop areas	Improved environmental qualities	

Building Name, Location, Date of Completion	Height (Architectural), Number of Floors Above Ground, and Gross Floor Area (GFA)	Green Retrofit Features Employed	Gained Benefits	Awards
		Upgraded 68 elevators	Reduced energy consumption and generated energy	
Willis Tower Chicago, Illinois, United States 1974	442 m/ 1,451 ft 108 stories 423,638 m²/ 4,560,001 ft²	Applying insulating thin-film to windows	To reduce heat transfer	Work in progress
		Upgrading the building's mechanical systems by supplying new gas boilers that utilize fuel cell technologies	To generate electricity, and support heating and cooling	
		Upgrading the building's 104 high-speed elevators and 15 escalators	To reduce energy consumption and generate energy	
		Installing advanced lighting control systems that automatically dim the lights and adjust to an optimal level of brightness when daylight is detected	To reduce energy consumption	

(Continued)

Table 1: Major or "Deep" Green Retrofit of Tall Buildings: Summary of Project Examples (Source: developed by the author) (*Continued*)

Building Name, Location, Date of Completion	Height (Architectural), Number of Floors Above Ground, and Gross Floor Area (GFA)	Green Retrofit Features Employed	Gained Benefits	Awards
		Upgrading the building's plumbing system and restrooms by installing low-flow water fixtures on toilets, urinals, and faucets	To reduce water consumption	
		Providing condensation recovery and irrigation systems	To reduce water consumption and collect "wasted" water	
		Integrating green roof as well as wind turbines and photovoltaic (PV) panels	To generate renewable energy	
Taipei 101 Taipei, Taiwan 2004	508 m/ 1,667 ft 101 stories 357,719 m²/ 3,850,455 ft²	Upgraded the electrical and mechanical systems of heating, ventilation, and air conditioning systems	Reduce energy consumption and improved air quality	LEED Platinum certification
		Enhanced its Siemens Apogee building management system	Improved structural stability	

Building Name, Location, Date of Completion	Height (Architectural), Number of Floors Above Ground, and Gross Floor Area (GFA)	Green Retrofit Features Employed	Gained Benefits	Awards
		Installed temperature and humidity sensors on each floor	Improved air quality	
		Added ice batteries	Reduced energy consumption needed to cool the building in summertime	
		Integrated an advanced irrigation system	Reduced water consumption	
		Added food-waste recycling system	Reduced generated waste	
Adobe World Headquarters (Three Towers) San Jose, California, United States 1996 1998 2003	West Tower: 79 m/259 ft 18 stories East Tower: 72 m/236 ft 16 stories Almaden Tower: 72 m/236 ft 18 stories 91,134 m^2/ 980,953 ft^2	Reduced the operating hours of the parking garage's exhaust fans from 24 to 6.	Reduced energy consumption	Platinum LEED certification

(Continued)

Table 1: Major or "Deep" Green Retrofit of Tall Buildings: Summary of Project Examples (Source: developed by the author) (*Continued*)

Building Name, Location, Date of Completion	Height (Architectural), Number of Floors Above Ground, and Gross Floor Area (GFA)	Green Retrofit Features Employed	Gained Benefits	Awards
		Provided a Web-Based IBIS that monitors and controls the building's subsystems by displaying electricity, water, gas, UPS systems, data centers and standby generators status in real-time.	Reduced consumption of water, gas, electricity, heating and cooling	
		Installed waterless urinals, automated flush valves, faucets, and soap and paper-towel dispensers	Reduced water consumption and improved environmental condition and cleanness	
		Relandscaped gardens with local and drought-tolerant plants	Reduced water and maintenance needs	
		Upgraded the subsurface drip irrigation system	Reduced water consumption	
		Upgraded chillers by installing an adaptable frequency drive (AFd)	Reduced energy consumption	
		Employed effective composting and recycling of paper, cardboard, plastic, glass, cans, printer toner, batteries, kitchen grease, and the like	Improved environmental conditions	

Building Name, Location, Date of Completion	Height (Architectural), Number of Floors Above Ground, and Gross Floor Area (GFA)	Green Retrofit Features Employed	Gained Benefits	Awards
		Installed locked bike cages and offered its employees incentives to use public transit	Improved environmental conditions	
Glastonbury House London, United Kingdom 1969	66 m/ 215 ft 23 stories	Installed dual flush toilets and spray taps	Reduced water waste	N.A.
		Replaced light fixtures	Increased lighting levels and saved energy	
		Installed renewable energy systems including solar PV and wind turbines	Reduced demand on city grid	
		Utilized eco-friendly materials	Improved microclimatic conditions	
The Joseph Vance Building Seattle, Washington, USA 1929	14 stories 1,2760 m²/ 137,350 ft²	Added automated sunshades	Reduced heat gain	LEED for EBs Gold certification

(Continued)

Table 1: Major or "Deep" Green Retrofit of Tall Buildings: Summary of Project Examples (Source: developed by the author) (*Continued*)

Building Name, Location, Date of Completion	Height (Architectural), Number of Floors Above Ground, and Gross Floor Area (GFA)	Green Retrofit Features Employed	Gained Benefits	Awards
		Recalibrated steam heating system	Reduced heat loss	
		Replaced ducted heating system	Improved air quality	
		Added localized thermostats	Enhanced system efficiency	
		Restored windows to the original design	Facilitated natural ventilation	
		Provided lighting retrofit with automated controls	Reduced energy consumption	
		Added light shelves	Reduced heat gain	
		Added CO_2 sensors	Improved safety	
		Replaced roof with a LEED-approved one	Reduced heat island effect	
		Replaced water fixture with T8 and T5 fixtures	Conserved water	

Building Name, Location, Date of Completion	Height (Architectural), Number of Floors Above Ground, and Gross Floor Area (GFA)	Green Retrofit Features Employed	Gained Benefits	Awards
		Provided bike storage and shower facilities	Supported green transport	
Hanwha Headquarters Seoul, Korea 1980	108 m/ 354 ft 21 stories 57,696 m²/ 621,035 ft²	Integrated a shading system into the façade	Reduced need for cooling interior space	CTBUH Renovation Award 2009 Regional Top 10 Awards", AIA Seattle
		When appropriate, angled glazing away from direct sunlight, while in other places angled glazing to receive direct sunlight	Reduced need for cooling interior space	
		Incorporated PV cells with the same angles	Harnessed solar energy	
		Remodeled several spaces (e.g., meeting rooms, auditorium, and executive areas) and incorporated sky gardens and vegetated lobbies	Improved livability	
		Improved interior climate and air quality	Improved tenants' health	
		Incorporated green walls	Improved tenants' health	

than keeping all the rooms at a comfortable 22°C (72°F) at all times, the introduced motion-sensing thermostats retain this ideal temperature only when the rooms are occupied. While the room is unoccupied, the system makes it go into "deep setback" by temporarily deactivating the cooling and heating system. However, once a guest checks in, the thermostat immediately resets the system and reactivates cooling or heating (as required) so that the room's temperature becomes comfortable by the time the guest reaches it. When the guest leaves the room (e.g., for meetings or shopping) the system goes into "deep setback" by temporarily deactivating the heating and cooling system to a threshold so that the system is able to bring the temperature back to the ideal within few minutes of the return of the guest. Finally, when the guest checks out, the system goes back into "deep setback" [46].

Other green renovation projects involve incorporating green roofs and walls and sky gardens. These "vertical landscaping" elements aim at improving air quality, ambiance, amenity space, visual quality, ecology, stormwater management, energy conservation, air cleaning, and mitigating urban heat island effects, to name a few. Green roofs and walls improve the building's thermal performance, provide longer roof membrane life span, enhance fire resistance and sound insulation, increase marketability, and improve the ability to turn the wasted roof space into various types of amenity space for building occupants. One early example of green roof retrofit projects is that of the 11-story City Hall of Chicago, completed in 2001. The planting palette features an extensive variety of plants, including native prairie and woodland grasses, forbs, and shrubs, hardy ornamental perennials, and grasses. An example of a green wall retrofit project is offered by the 18-story Portland's Edith Green/ Wendell Wyatt Federal Building. The 76 m (250 ft) west wall lets plants grow on "fins" or vertical trellises. During spring and summer, greeneries shade the façade and hence reduce the energy required to cool the building. In the winter, the plants naturally drop their foliage, and sunlight enters the building to warm interior spaces. Captured rainwater from the roof mainly irrigates the green wall and is supplemented by municipal water when necessary [47,48].

6.0   Continuing Challenges

Research indicates that the greatest barriers to green retrofits include lacking robust cost–benefit equations, adequate financial and technical support, tenant's behavior, split incentive issue, operation and management, and no one template to apply. Historic buildings also face additional challenges. These issues are detailed as follows [15,18,22–26].

6.1   Cost-Effectiveness

Proving the cost–benefit equation is key to convince building' owners to commit to a green retrofit. To make an investment, an owner wants proven methods and technologies and demands concrete ideas of the ROI. Often, the energy performance prediction tools are imperfect and every building responds differently. Consequently, there is a sense of lack of confidence about the cost-effectiveness of undertaking a green retrofit. Another potential barrier related to cost-effectiveness could be the amount of time it takes to obtain cost savings from renovations. In many cases, the time needed is several years. Indeed, there are instances where a retrofit is less cost-effective, and new construction will be more effective and desirable by owners and developers.

6.2   Financial Resources

An owner may be convinced by the presented cost-effective analysis and persuaded by the retrofit's merits but may lack financial resources. Therefore, securing adequate funds for green retrofit is often a barrier. Buildings' owners are looking for a quick ROI in 2–4 years; and for a major retrofit, it is likely going to take longer than that. Further, in urban areas, there is a financial incentive to maximize the use of sites by adding floor spaces to achieve economies of scale and height for views, which translate in increasing rent values. However, these actions are far from being attainable in an EB in urban cores. Also, developers often perceive little economic justification for retaining EBs and instead look for developable land rather than retrofitting EBs [26].

6.3 Codes and Regulations

Building policies and codes, in general, form a major obstacle to a building retrofit project. For example, building codes in the United States have historically favored new construction over retrofit. In the absence of flexible land use regulations and incentives for reuse, older buildings are commonly torn down to make way for larger structures. Energy codes can also sometimes deter building reuse, as they are typically not well-adapted to the unique limitations and opportunities presented by individual buildings. When these issues are added to seismic and Americans with Disabilities Act (ADA) requirements, they collectively form the "tipping point" in decisions favoring demolition.

6.4 Split Incentives

Projects are challenged by the split incentive problem, a phenomenon that refers to the fact that a building's owner pays for building efficiency while tenants reap the rewards. In many instances, building owners delay or avoid making efficiency investments because it is only their tenants—those paying the utility bills—reap the financial benefits. Simultaneously, tenants are often hesitant to invest in energy upgrades on properties they do not own.

6.5 Technical Challenges

Overall, required technical expertise on the part of the project team—architects, engineers, building managers, tenants, and energy service companies—continues to be lacking. Additionally, retrofit work is often regarded as riskier than constructing new buildings because the process can be less predictable, and many developers fear unforeseen technical challenges once rehabilitation is underway.

6.6   *Tenants' Agreement*

In some cases, tenants are the most challenging factor. Once tenants reside in a building, an owner needs to obtain their permission to retrofit the building. The owner also should ensure minimum interruption and inconveniences imposed on tenants while the building goes under retrofit. Further, building systems, such as lighting and HVAC, should remain functional and operational during a retrofit.

6.7   *Operation and Management*

In order to guard the sustainability of a retrofit project, building owners and managers should ensure a match between the provided systems and tenants' behavior. Research and experience confirm that the building's performance degrades when there is a mismatch between these two. There have been some incidences where buildings equipped with the best technologies perform the worst because of mismanagement. For example, the override feature in a system could be useful to meet temporal needs or respond to an emergency situation. However, it is important that managers reset the system back to normal. Otherwise, tenants may experience discomfort and they will likely blame the new system and equipment, not the improper management of the system.

6.8   *No One Template*

Due to vast variations of buildings' conditions as well as variations in geography, climate, culture, and finance, it is near impossible to develop a working template that developers can use and replicate elsewhere. Developers, architects, and engineers need to perform substantial studies before embarking on an effective retrofit. Certainly, basic familiarity with a region, climate, culture, and local construction practices may shed insight on the process [2,24].

6.9 *Historic Buildings*

Retrofitting older and historic buildings particularly of large-scale (the case of tall buildings) faces challenges and possible compromises as the design team balances energy performance with a plethora of competing priorities. Environmentalists and preservations often get in conflict when the appearance of a historic building needs to be changed for the sake of improving the building's energy performance. The complications on a project may result from historic preservation standards that often care about retaining historic features, fabric, and character even at the fine-grain scale. Further, some buildings could be treated as "historic" although they do not necessarily acquire a formal recognition of being a historic landmark. For example, when green retrofit was proposed to the Willis Tower (formerly Sears Tower), many denizens voiced objections fearing that changes may impact its perceived iconic image. Conducting such retrofit would be a challenge for providing environmental benefits, while retaining the image of the building unchanged [24].

6.10 *Safety*

A retrofit project engenders a stream of activities that may endanger the building's tenants and pedestrians. For example, projects that involve façade retrofits may entail the accidental dropping of materials and tools on pedestrians. Similarly, projects that involve retrofitting interior spaces may disturb tenants, and accidents could happen as well.

6.11 *Security*

Finally, some buildings, e.g., government buildings, could require high-security and conducting retrofit projects in these buildings would be challenging [25,26]. Interestingly, some buildings' retrofit projects may face a combination of the aforementioned challenges. For example, the Dirksen Federal Building completed in 1964 was recently undergone a green retrofit project that faced many challenges including "asbestos remediation, historic preservation standards, maintenance of high-security levels and, all-glass transparent facades, most of all, conducting work in and around the occupants of an operating courthouse" [26].

Consequently, researchers, architects, engineers, and city officials should collaborate on easing the above challenges. This endeavor can be stressed and appreciated when we learn that the alternative of a retrofit, could be a costly and "unsafe" demolition. Commonly used demolition methods, such as implosion or wrecking ball, are often rendered to be unsafe. As mentioned earlier, in the case of demolishing the Royal Canberra Hospital in Australia, the explosive demolition resulted in injuring nine people and killing a child, who was struck by flying debris. Besides, neighboring tenants, pedestrians, and properties suffer from excessive dust, vibration, and noise. Similarly, potential dropping materials from upper floors using nonexplosive methods (e.g., piece-by-piece dismantling) is a serious problem. In addition to tenants and pedestrians, safety concerns include site workers, supervisors, operators, and engineers due to potential hazards resulting from flammable materials. All these safety issues become more serious in denser and crowded urban environments, often the homes of tall and supertall buildings.

7.0   Conclusion

This chapter offers a glimpse of green retrofit projects. Since buildings consume a significant amount of energy and because EBs comprise the largest segment of the built environment, it is important to initiate energy conservation retrofits to reduce the need for heating, cooling, and lighting buildings. Careful material selection and efficient design strategies for reuse can play a major role in minimizing the impacts associated with building renovation and retrofit projects. Cautious retrofits should reduce operational costs and environmental impacts, and increase building adaptability, durability, and resiliency.

A successful retrofit requires applying an integrated, whole-building design process. The integrated project team should conduct a preliminary study that considers the various elements that need retrofits. Then, the team may select fewer design strategies that will help to achieve the project's top objectives while yielding maximum benefits. As such, "focused" strategies should lead to optimal saving of resources, substantial improvement of property value, and a considerable increase in building longevity. The retrofit project should also provide a better, healthier, and more comfortable environment for people to live, work, and play. Finally, it is important to consider the building age and pick the right time for a deep retrofit rather than being motivated by purely financial incentives offered by governments and environmental agencies. A major retrofit must address already existing needs and operational problems. Otherwise, a green retrofit will not be cost-effective.

Much of the examined case studies in this chapter inform and inspire about green retrofit processes. However, as mentioned earlier, given the differing building conditions, ages, geographic locations, climatic conditions, building materials, and financial resources, it is near impossible to create a "blue template" for a green retrofit. Instead, examining case studies, as this chapter attempts to do, is the starting point to learn about the subject matter. Fortunately, accumulative experiences will certainly lead to identifying and embracing more efficient ways to retrofit aging buildings.

8.0 Future Research

Undoubtedly, this research is incomplete. Future research may conduct quantitative assessments of the involved costs and benefits of green retrofit projects. For example, it would be useful to compare the building's performance on energy consumption before and after the retrofit. To do so, the researchers should have access to energy and utility bills before and after the retrofit. The author could not acquire access to such data, and we hope that building owners and utility companies would be more cooperative with university researchers. In the same token, future research may engage architectural and engineering firms that conduct building retrofits to obtain some detailed information about technical issues, the involved process, and incurred costs.

References

(1)   Department of Energy. https://www.energy.gov/. Accessed March 1, 2020.

(2)   U.S. Green Building Council. https://www.usgbc.org/. Accessed March 1, 2020.

(3)   Al-Kodmany K. *The Vertical City: A Sustainable Development Model*. Southampton, UK: WIT Press; 2018.

(4)   U.S. Energy Information Administration. https://www.eia.gov/. Accessed March 1, 2020.

(5)   EnergyStar.Building upgrade manual. 2016. https://www.energystar.gov/buildings/tools-and-resources/building-upgrade-manual. Accessed March 1, 2020.

(6)   Al-Kodmany K. Green retrofitting skyscrapers: a review. *Build*. 2014;4(4):683–710. https://doi.org/10.3390/buildings4040683. Accessed March 1, 2020.

(7)   Samuel S, Mehtab N. *Revisiting the Case of Sustainable Construction via LCA—Build New or Reuse?* Michigan, MI: SERF Foundation; 2014.

(8)   Gutierrez D. How do green building principles support energy efficiency? *Straughan Environmental*. 2019. https://www.straughanenvironmental.com/how-do-green-building-principles-support-energy-efficiency/. Accessed March 1, 2020.

(9)   Kahn K. Iconic skyscrapers invigorated by going green. *USA Today*. July 4, 2009. http://usatoday30.usatoday.com/news/nation/2009-07-04-skyscrapers-green_N.htm. Accessed March 1, 2020.

(10) Roos G. Green' buildings a mixed bag as budgets contract. *Environmental Leader*. August 3, 2009. http://www.environmentalleader.com/2009/08/03/green-skyscrapers-earn-higher-lease-rates-utility-projects-scale-back/. Accessed March 1, 2020.

(11) Campbell-Dollaghan K. Tokyo will retrofit its skyscrapers to prep for the next big quake. *Gizmodo*. 2013. http://gizmodo.com/tokyo-will-retrofit-its-skyscrapers-to-prep- for-the-nex-981057523. Accessed March 1, 2020

(12) Dupre J. *Skyscrapers: A History of the World's Most Extraordinary Buildings*. New York, NY: Blackdog and Leventhal Publishers; 2013.

(13) Paradis R. Retrofitting existing buildings to improve sustainability and energy performance. *Whole Building Design Guide*. 2016. https://www.wbdg.org/resources/retrofitting-existing-buildings-improve-sustainability-and-energy-performance. Accessed March 1, 2014.

(14) Stanfield R. The Sheraton Chicago hotel & towers: greening Chicago's skyline with committed owners, innovative technology and two very smart guys in the basement. 2012. https://www.nrdc.org/experts/sheraton-chicago-hotel-towers-greening-chicagos-skyline-committed-owners-innovative. Accessed March 1, 2020.

(15) Al-Kodman K. The sustainable city: practical planning and design approaches. J. Urban Technol.2018;25(4):95–100.

(16) Lai J, Wang S, Schottler M, Mahin S. *Seismic Evaluation and Retrofit of Existing Tall Buildings in California*. London, UK: Pacific Earthquake Engineering Research Center; 2015.

(17) Paradis R. Retrofitting existing buildings to improve sustainability and energy performance. In: *Whole Building Design Guide*. Washington, DC: National Institute of Building Sciences; 2012. http://www.wbdg.org/resources/retro_sustperf.php?r=sustainable. Accessed June 1, 2014.

(18) Castleton HF, Stovin V, Beck SBM, Davison JB. Green roofs; building energy savings and the potential for retrofit. *Energy Build*. 2010;42(10):1582–1591.

(19) Lepik A. *Skyscrapers*. Munich, Germany;: Prestel Verlag; 2008.

(20) Rocky Mountain Institute. *Practice Guide: The Path to a Deep Energy Retrofit Using an Energy Savings Performance Contract*. Rocky Mountain Institute: Snowmass, CO; 2015. https://rmi.org/insight/practice-guide-the-path-to-a-deep-energy-retrofit-using-an-energy-savings-performance-contract/. Accessed March 1, 2020.

(21) Coan S. Designing the city of the future and the pursuit of happiness. *Rocky Mountain Institute*; 2019. https://rmi.org/designing-the-city-of-the-future-and-the-pursuit-of-happiness/. Accessed March 1, 2020.

(22) Kok N, Miller NG, Morris P. The economics of green retrofits. *J Sustain Real Estate*. 2012;4:4–22.

(23) Matus M. The retrofit Chicago challenge gets skyscrapers to go green. *Inhabitat*. 2012. http://inhabitat.com/the-retrofit-chicago-challenge-gets-skyscrapers-to-go-green/. Accessed June 1, 2014.

(24) Al-Kodmany K. Green towers and iconic design: cases from three. *Int. J. Archit. Res.* 2014;8(4):11–28.

(25) Kalanki A. *Transforming the Global Comfort Cooling Market: China's Opportunity for Economic and Climate Leadership.* Rocky Mountain Institute; 2019.

(26) Gonchar J. Continuing education: seismic design architectural records continuing education center. 2019. https://www.architecturalrecord.com/articles/13874-continuing-education-seismic-design. Accessed March 1, 2020.

(27) Wells M. Skyscrapers: Structure and Design. London, UK: Laurence King Publishing; 2005.

(28) Yeang, K. Ecoskyscrapers and ecomimesis: new tall building typologies, in Wood, A. (Ed.): *Proceedings of the 8th CTBUH World Cong. on Tall & Green: Typology for a Sustainable Urban Future, CD-ROM*, 84–94; 2008.

(29) Wood A. Best Tall Buildings 2012, CTBUH International Award Winning Projects, Council on Tall Buildings and Urban Habitat (CTBUH). New York and London: Routledge, Taylor & Francis Group; 2012.

(30) Al-Kodmany K. Sustainability and the 21st century vertical city: a review of design approaches of tall buildings. *Build.* 2018;8(8):44

(31) Al-Kodmany K. Tall buildings and elevators: a review of recent technological advances. *Build.* 2015;5(3):1070–1104. https://doi.org/10.3390/buildings5031070. Accessed March 1, 2020

(32) J.C.S. Goncalves. The Environmental Performance of Tall Buildings .London, UK: Earth Scan; 2010. CTBUH.org/paper. Accessed March 1, 2020.

(33) Hargreaves S. Empire state building cuts energy use 20%. *CNN Money.* May 7, 2012. http://money.cnn.com/2012/05/07/news/economy/empire_state_building/. Accessed June 1, 2014.

(34) Ng E. Designing for daylight, In: E. Ng (ed.)Designing High Density Cities for Social and Environmental Sustainability, Chapter 13. London, UK: Routledge, Earthscan: 2010;181–291.

(35) Chandler J. Why does LA's mandatory retrofit program ignore vulnerable steel skyscrapers. *Curbed.* January 17, 2019. https://la.curbed.com/2019/1/17/18179326/earthquake-skyscrapers-steel-los-angeles. Accessed March 1, 2020.

(36) Al-Kodmany K. Sustainable tall buildings: cases from the Global South. *Int. J. Archit. Res.* 2016;10(2):52–66. CTBUH.org/paper. Accessed March 1, 2020.

(37) Leung B. Greening existing buildings [GEB] strategies. *Energy Rep.* 2018;4:159–206. doi:10.1016/j.egyr.2018.01.003.

(38) Al-Kodmany K. Skyscrapers in the twenty-first century city: a Global snapshot. *Building.* 2018; 8(12):175.

(39) Knox RH. Case study: Adobe's "Greenest Office in America" sets the bar for corporate environmentalism. *FMLink.* 2015. http://www.fmlink.com/article.cgi?type= Sustainability&pub=USGBC&id=40625&mode=source. Accessed March 1, 2020.

(40) Cammell A. Adobe systems' green initiatives generate huge savings. *Tradeline.* July 29, 2008. http://www.tradelineinc.com/ reports/2008-7/adobe-systems-green-initiatives- generate-huge-savings. Accessed on March 1, 2020.

(41) Higgins C. *A Search for Deep Energy Savings* (Final report). Vancouver, WA, Canada: New Buildings Institute; 2011. https:// newbuildings.org/sites/default/files/NEEA_Meta_Report_Deep_ Savings_NBI_Final8152011.pdf. Accessed March 1, 2020.

(42) Green D. My zero energy retrofit beats my 401(k). *Zero Energy Project.* September 23, 2018. https://zeroenergyproject. org/2018/09/23/my-zero-energy-retrofit-beats-my-401k/. Accessed March 1, 2014.

(43) Frearson A. UNStudio's dynamic facade will control indoor climate for Seoul tower, dezeen. April, 2014. https://www. dezeen.com/2014/04/30/hanwha-hq-seoul-unstudio-dynamic-facade/. Accessed March 1, 2020.

(44) Oldfield P. *The Sustainable Tall Buildings: A Design Primer.* London, England: Routledge; 2019.

(45) Al-Kodmany K. *Eco-Towers: Sustainable Cities in the Sky.* Southampton, UK: WIT Press; 2015.

(46) Slowey K. NYC's 'Green New Deal' to ban glass, steel. *ConstructionDive.* April 24, 2019. https://www.constructiondive. com/news/nycs-green-new-deal-to-ban-glass-steel-skyscrapers/553323/.

(47) Al-Kodmany K, Ali M. *The Future of the City: Tall Buildings and Urban Design.* Southampton, UK: WIT Press; 2013.

(48) Domonoske C. To fight climate change, New York City will push skyscrapers to slash emissions, NPR, WBEZ91.5. 2019. https:// www.npr.org/2019/04/23/716284808/new-york-city-lawmakers-pass-landmark-climate-measure. Accessed March 1, 2020.

CHAPTER 6
Energy Efficiency as a Function of Changing Building Form in Sustainable Rural Architecture

Şefika Ergin[1], Kemal Çorapçıoğlu[2], Figen Balo[3]
[1]Architecture Department, Dicle University
[2]Architecture Department, Mimar Sinan Fine Arts University
[3]Industrial Engineering Department, Firat University

Abstract

Each type of rural building example has different materials and construction techniques based on the geographic and topographic conditions of their location. This generates unique architectural identities that are also shaped by cultural norms of the region. While rural buildings are in the same climate zone, they can have different planning types and configurations. Different types of building form can also be seen in close settlements. This diversity is born of optimal adaptation to climatic comfort conditions, depending on the differences in construction materials. For this reason, the most effective parameters in evaluating performance differences are a range of housing types, space organizations, building forms, construction systems, and material types used. Conventional architecture would suggest that efficient design thinking creates appropriate climatic conditions without utilizing excessive power. The building forms' thermal efficiency, which is produced from traditional architecture, is assessed within the characterized settlement textures and the conclusions existing.

The interplay between energy loads, building form, and settlement texture will help to drive the evolved architecture that will enhance efficiency even further.

This paper proposes to investigate the settlement texture and building form's impact on energy efficiency by analyzing the project characteristics of traditional Diyarbakır, Turkey, rural area buildings. This will be conducted through the Ecotect simulation program. The quality and quantity of the add-on architectural plans in rural houses in Diyarbakır were analyzed in terms of their energy efficiency. The intent is to preserve and maintain rural architectural identity elements shaped by the data obtained from the physical and natural environment. In this way, two exemplary rural architectures have been studied to educate designers in the ways of livable and energy-efficient space creation— while maintaining a sustainable and traditional rural culture. The buildings' outer surface area and volume were increased incrementally. The first of the house types investigated was enlarged in three stages and the second in four stages. Comparing the A/V ratio in the first stage and the area/volume ratio in the last stage, House1 and House2 achieved 42.93% and 75.96% energy efficiency at life space, respectively.

Keywords: Energy efficiency, building form variation, sustainable architecture, traditional buildings, rural architecture

1.0 Introduction

Rural settlements were formed over many centuries as a result of the sociocultural structure of the society and an organic development shaped by integrating with nature. Local structures, which are the main elements of rural settlements, constitute the original settlement texture with variable factors such as social and cultural lifestyle of local people, livelihoods, climatic factors, topography, and materials. The factors that make up the settlement texture are effective in shaping the architectural environment in rural areas by gaining quality according to the physical environment characteristics. The local architectural formation, which varies from settlement to settlement, reflects the architectural identity of the region. Architectural identity formation in rural settlements is the most important element of that geography. It is important to ensure the sustainability of the original textures and architectural identity of the local settlements in order not to lose social and cultural values.

Furthermore, the most important element that determines the architectural identity in rural settlements is residential buildings. The housing, which meets the housing and living needs of the household, is also an important factor in maintaining the local architectural identity. Depending on the qualitative and quantitative parameters of the building, it is important to meet the performance criteria that it will display against physical environmental conditions in determining the level of comfort that must be realized within the structure. The structure that performs the best in physical environmental conditions is the one that provides suitable and comfort conditions in the spaces and reduces the need for heating, ventilation, air conditioning, or lighting of the building with any energy system, as well. From this perspective, evaluating and designing the plans with physical environmental conditions is important in terms of providing the interior climate, which is comfort in the spaces of the building.

Moreover, climatic comfort is the ability to achieve the desired level of physical and mental comfort by consuming minimum energy in the area. According to ASHRAE Standard 55-2004, climatic comfort is defined as the conditions in which a person is satisfied with his climatic environment. Since the physiological and psychological differences differ from person to person, the environmental conditions required for comfort are not the same for everyone. The goal of ensuring optimal conditions should be to provide environmental conditions that satisfy as many people as possible. It is possible to define the conditions in which 80% or more of users who find their environment acceptable in terms of climatic comfort as climatic comfort conditions [1,2]. The parameters related to the building in providing climatic comfort can be defined as location selection, building intervals, building form, orientation of the building, space organization, material, and construction technique. These parameters play an important role in the formation of indoor climate conditions. The minimum usage of heating and cooling, and making systems used to provide climatic comfort in buildings is important in terms of energy consumption. The exhaustion of energy resources in today's conditions necessitates the optimum use of natural energy resources in terms of energy saving. The design, planning, and layout of the buildings in harmony with the physical and environmental conditions, can contribute to energy efficiency by providing optimum gain in terms of the energy used. For this reason, energy-efficient building design requires the creation of active and passive control facilities suitable for the building and determining the design criteria for increasing the performance of the building in terms of heating, cooling, ventilation, lighting, and providing energy conservation.

Additionally, energy performance of buildings (in terms of indoor comfort conditions, fossil-based fuel consumption, harmful emissions, etc.) only to individual building components (interior and exterior walls, windows, floors, etc.) or plumbing system (heating, ventilation, air conditioning, lighting, etc.), is not based on their dynamic interaction as an integrated whole [3–5]. Several heat sources operating through various kinds of energy and mass transfer paths can determine building indoor conditions [6–8]. The main sources affecting mentioned above are

- The most important variables: outdoor climate with air temperature, humidity, solar radiation, and wind speed and direction
- Users that cause metabolic heat gain: various equipment, artificial lighting, etc.
- Auxiliary equipment, which are devices used for heating, cooling, and/or ventilation

These sources are effective in indoor conditions with the help of various energy and mass transfer processes listed below [8]:

- Conduction through the building envelope and interior divider walls

- Solar radiation passing through the transparent surfaces of the building envelope and reradiation by transforming into long-wave radiation on the surfaces
- Convection for heat exchange between surfaces and air
- Air movement through the building envelope (controlled-natural ventilation; uncontrolled infiltration/exfiltration) inside the building and with the heating, cooling, and ventilation system
- The movement of fluids in the plumbing system

Performance-based building energy regulations require determining the maximum acceptable energy consumption levels without defining the method, material, and process. Performance-based compliance requires studying and evaluating consumption levels based on the integrated performance of building elements and systems such as building envelope, lighting, and air conditioning. However, determining the extent of the assessment depends on the purpose and limits of the assessment [9–11].

The target of the energy performance building plan is to obtain the utilizers required by utilizing less amount of energy. To obtain this target, energy performance plan characteristics should be thought of during the plan step with respect to ecological circumstances and, specifically, the climate, which is an effective element that proposes sufficient information to improve a congruent plan. In this scenario, different instances of traditional sustainable rural architecture supply that choose proper values for design characteristics thinking climate circumstances conclusions in comfort, indoor thermal environment levels without utilizing extreme energy for climatic conditions. As a result, buildings of traditional sustainable rural architecture should be valued as the primary resource of information gained by many years' practice and contemporary construction of modern buildings that may advantage from this inheritance. However, this information is ignored and buildings are built without taking environmental considerations into consideration today. For example, Turkey has five distinct climatic regions and each of the regions of the environment offers competent specimens of deep-rooted conventional architecture, but modern structures were designed on the basis of a standard template, which demands a large volume of energy to offer warmth due to their insufficient construction. On the other side, according to the World Bank's study of climate change and climate details, the region of the hot-dry environment zone will increase due to the global warming's desertification impact and Turkey will be seriously affected by 2050 [12]. Of this purpose, construction criteria of energy efficiency should be tested in order to address the anticipated adjustments and to plan to build in accordance with the usual extreme climate conditions.

Because of the sequence of the days and seasonal variations, the thermal behavior of the climate is quite distinctive in Diyarbakır province. The summer days are usually very hot and dry but the nights are cool [13]. In Diyarbakır province, these extreme situations push the design to consider simple and climate-friendly ways to provide safe indoor environments. This research is aiming to explore the energy performance of conventional settlements according to the changing building form. For this purpose, two different buildings in rural architecture were analyzed in terms of energy performance according to the change stages in the building form. The outer surface area and volume of the buildings have been gradually increased with the areas added to these buildings. This increase was realized in line with the creation of a new building configuration, which is based on the area/volume (A/V) ratio of the buildings. In this way, the effect of the changed stages in the building form on reducing the annual energy need in the main living area was assessed. Currently, there are many studies aiming to calculate the annual total energy amount of the current state of any building. In addition, there are no studies analyzing the annual total needed energy of a specific structure over time in the literature, depending on the needs and requirements. In this region, this study is important because it is an exemplary study to analyze how the heating and cooling loads, and total energy needs of the building change according to the A/V ratio of the building.

2.0   General Properties of Diyarbakır Province in View of Analysis

2.1   Workspace and Features

The Karacadag town in Diyarbakır province was chosen as the study area. The map of Diyarbakir province is displayed in Fig. 1. Diyarbakır province, located in the Southeastern Anatolia Region, is located at coordinates 37° 55' north latitude and 40° 12' east longitude. Its height above sea level is 670 m. There are Siirt and Muş in the east of Diyarbakır; Mardin in the south; Sanlıurfa, Adiyaman, and Malatya in the west; and Elazig and Bingol in the north. Karacadag is on the eastern edge of the wide basalt plateau that extends from the extinct volcano mass to the Tigris river. It is located in a pit area surrounded by mountains on a horizontal surface that is 160 m above the Tigris valley [14].

There are the Taurus Mountains to the north, Karacadağ to the west, and Mardin hills to the south. With its position between the mountain plateaus on the north and the steppe plains in the south, it constitutes a transition area suitable for settlement [15]. In Diyarbakır province, which has an area of 15,354 km², the Tigris River is the only river that irrigates these lands with its various tributaries.

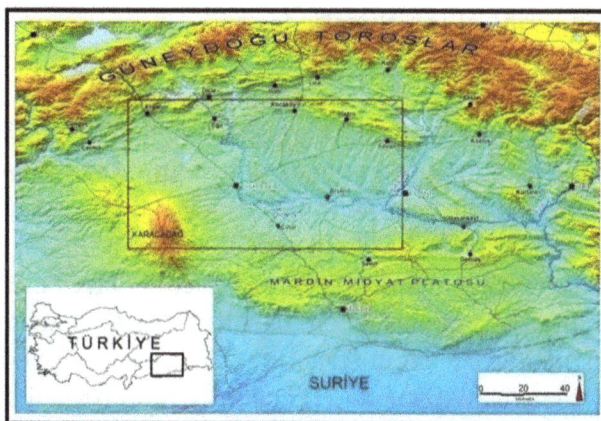

Fig. 1: The Map of Diyarbakir Province

2.2 General Features of Rural Architecture of Diyarbakır Province

It has been understood that the social life of the local people is a very effective factor in architectural shaping during the examinations made in the region and in one-on-one interviews with the building users. The usage of the local people in the hot and cold periods of the year may differ according to their original living cultures. Residential areas, which are used extensively in the cold periods of the year, are used less frequently in the warm periods of the year. During the warmer periods of the year, most of the households' activities such as eating and sleeping are carried out outside the home. In hot periods, the most active function is the sofa. Most of the eating, gathering, and sleeping activities take place on the sofas. The use of the sofa in planning types in residential architecture consisted of examples with front sofas, side sofas, L sofas, and medium sofas. Among the examined structures, the use of the middle sofa is common, especially in single building types. There are connections with the spaces in the two opposite sides of the hall and one door opening to the outside in the other opposite sides in the middle table plan types. These two doors, which are kept open during the hot summer months of the year, provide a constant airflow in the sofa area and create a natural air conditioning effect in the space. It was observed that the use of this sofa in the planning provided an advantage in terms of the climatic features of the region during warm periods. It is seen that the quarries built for the purposes of cooking or baking bread in the residences in the region were designed within the structure. Quarries, which are generally planned in the sofa area or kitchen, contribute to the thermal gain of the place during the cold periods of the year.

In the measurements made in the buildings in the region, the thickness of the walls has been observed to vary between 55 cm and 80 cm in structures where basalt material is used, and between 55 cm and 75 cm in structures where limestone material is used, and between 40 cm and 60 cm in structures where adobe material is used. Average wall thickness: 0.68 m in basalt walls,

0.66 m in limestone walls, and 0.50 m in mudbrick walls. It was measured that the outer wall thicknesses of some buildings in the region were higher than the inner wall thicknesses. This detailing has been made to reduce the impact of climatic conditions and to reduce heat losses from external walls as well [16].

Increasing the thickness of the outer walls allows wall heat transfer coefficient to be reduced, thereby increasing the heat transfer resistance. With this detailing, the formation and protection of indoor climate conditions are provided with less energy use. This application is effective in providing interior climatic comfort in the spaces. It is thought that applying these details in new constructions will be beneficial in terms of energy saving.

In the facade analysis, it was observed that the southern direction was preferred in order to get the most benefit from the sun. However, in complex planning, the direction preference is limited as the growth is realized by adding the structures to each other. In this case, the most appropriate direction was chosen for the current location. It has been observed that building users have developed solutions to reduce the outer wall surface area on the facades facing north. By placing the poultry houses of the household adjacent to this facade, the outer wall surface areas are reduced. It is recommended to reduce the outer wall surface area with such solutions and to prefer the indoor climate of the house where the house is adjacent in cold periods. It is recommended to use service units in the north direction or to make solutions to reduce the outer surface areas of the spaces on this front.

2.3   *The Climatic Features of Diyarbakır Province*

According to TS 825, provinces and some districts are classified as five different degree-day regions according to the climatic conditions of their geographical location. Turkey's climate zones by TS 825 are displayed in Fig. 2 [17]. In the classification made as degree-day regions according to provinces, Diyarbakır province is within the second Region degree-day provinces.

Fig. 2: Turkey's Climate Zones by TS 825 [17]

The average annual temperature values between 1929 and 2018 in Diyarbakır province are shown in Fig. 3 [18]. The average temperature, the average highest temperature, and the average lowest temperature values are obtained as 31.1°C, 38.3°C, and –2.3°C. The annual areal precipitation in Diyarbakır between 1981 and 2014 are displayed in Fig. 4 [18,19]. As seen in Fig. 4, the long-term average meteorological data of the region show that the province is under the influence of the hot-dry climate.

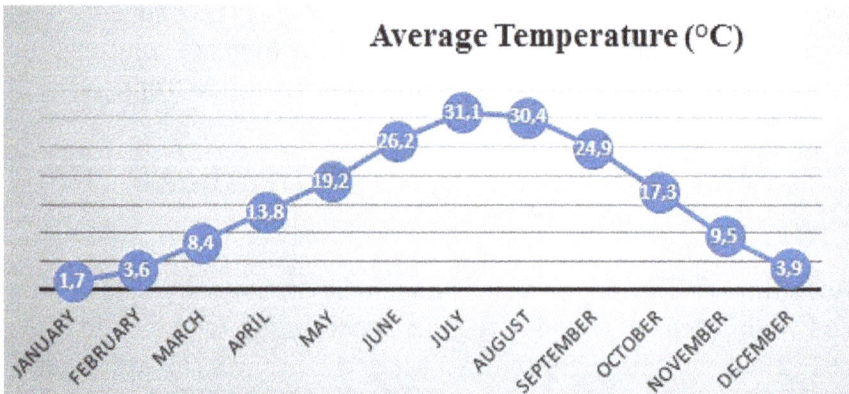

Fig. 3: The Average Annual Temperature Values Between 1929 and 2018 in Diyarbakır Province

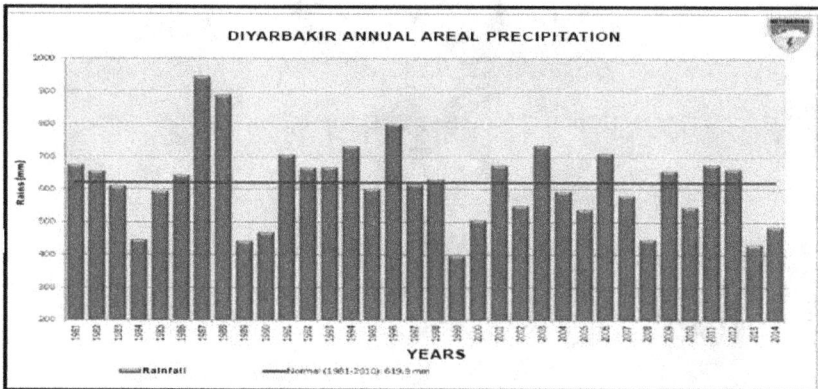

Fig. 4: The Annual Areal Precipitation in the Diyarbakır Province (1981–2014) [18,19]

The summer months are very hot and dry, and the winter months are cold and rainy. It has low cloudiness and low relative humidity. On the other hand, in summer and winter, temperature differences between day and night are high. The amount of solar radiation received by the province of Diyarbakır is quite high because it is located in the middle latitudes and spread over a flat topography.

Due to the prolongation of the days in the summer and the rays of the sun that reach the surface with steeper angles, the solar radiation intensity reaches 3.05 MJ/m² at noon in June and the daily sunbathing times are close to 12.4 hours daily in July. Since these values are high, solar radiation negatively affects comfort conditions in the summer.

Typical yearly data from the Diyarbakır province have been loaded into the Ecotect software, which is the simulation program that will be used in the analyses to be made. According to these data, the annual comfort diagram and the coldest day of the year, January 15, and the hottest day of the year, July 29, were obtained with the help of the simulation program.

Thermal comfort limit values of the Diyarbakır province by months and climatic information of the coldest day (January 15)

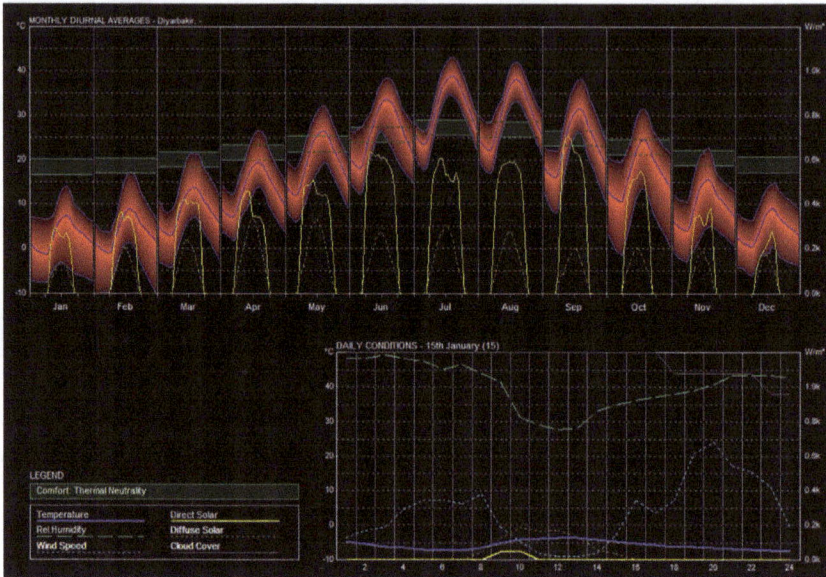

Fig. 5: Thermal Comfort Limit Values of Diyarbakır by Months and Climatic Information About the Coldest Day of the Year (January 15)

and hottest day (July 29) of the year are given in Figs. 5 and 6, respectively.

In the summer and winter months, the dominant wind speeds and directions in Diyarbakır are shown in Fig. 7. While Diyarbakır is open to the desert winds coming from the south due to its topographic location, it is located in an area closed to the cold winds coming from this direction thanks to the mountain ranges in the north. Looking at the wind directions, it could be seen that the caramel blowing in the northwest direction in winter is the dominant wind in this region [18].

The meteorological values determined by the Ecotect simulation program during the year of analysis applied in Diyarbakir province are given in Fig. 8. In Fig. 8, annual average temperature distribution, annual average humidity distribution, annual solar radiation distribution, and annual sky coverage rate are displayed, respectively. As seen in this figure, the hottest period of the year is June-July-August and the humidity level

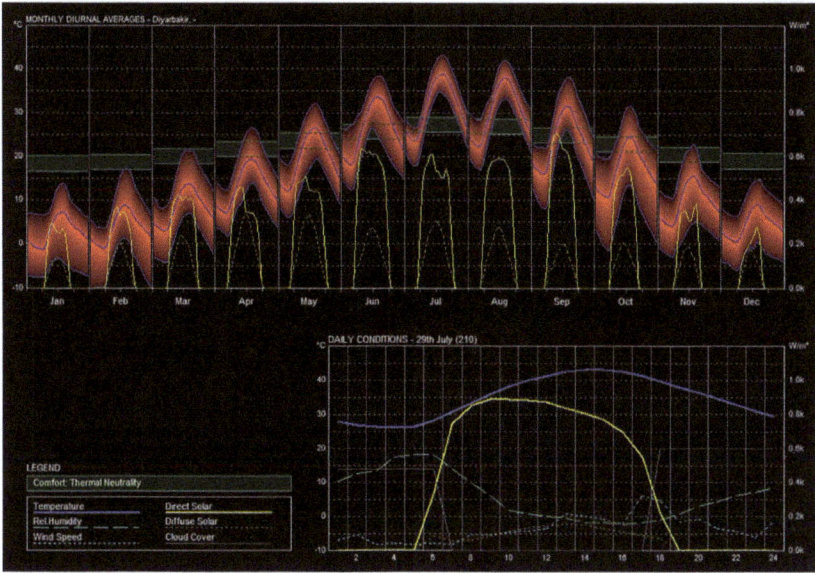

Fig. 6: Thermal Comfort Limit Values of Diyarbakır by Months and Climatic Information About the Hottest Day of the Year (July 29)

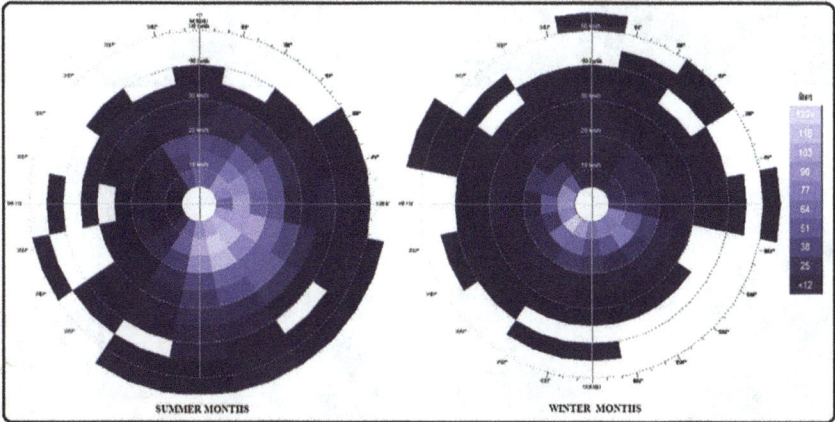

Fig. 7: The Dominant Wind Speeds and Directions in Diyarbakır

reaches the lowest level in these months. In these months, it is observed that the solar radiation distribution is high and the sky coverage rate is the lowest.

Fig. 8: Meteorological Values Determined by the Ecotect Simulation Program During the Year of Analysis Applied in Diyarbakir Province **(a)** Annual Average Temperature Distribution **(b)** Annual Average Humidity Distribution **(c)** Annual Solar Radiation Distribution **(d)** Annual Sky Coverage Rate

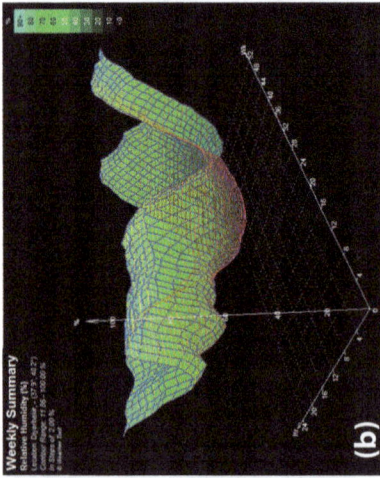

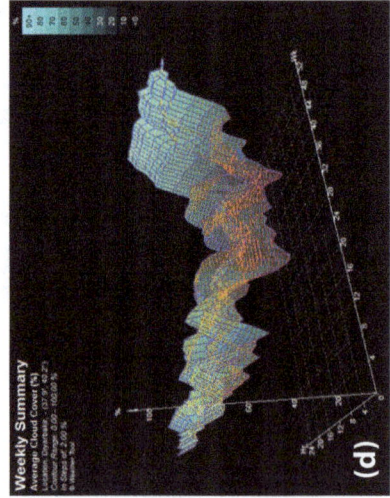

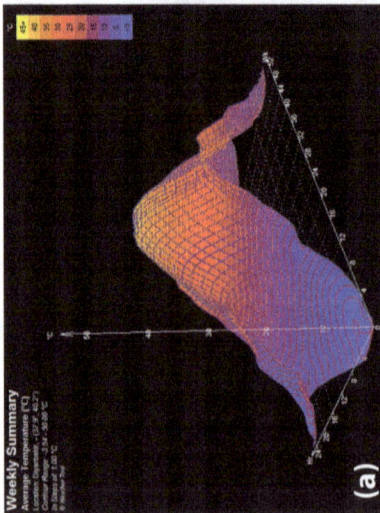

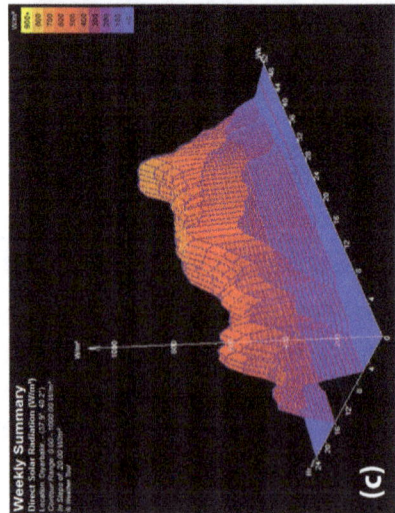

3.0   Analyzed Buildings in Diyarbakir Province

3.1   Architectural Features of the Analyzed Buildings

In this study, two residential buildings in Akdere rural settlement of the Silvan district of Diyarbakır province are discussed. These two buildings, which have a similar plan layout, underwent form change with the addition of the space made at different times. These structures are coded as House1 and House2. The location of the working structures is given in Fig. 9.

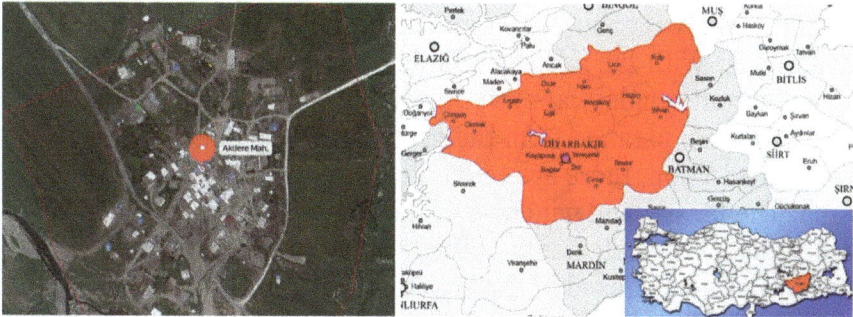

Fig. 9: Location of the Working Structures

3.1.1   The Plan Properties and Change Phases of House1

The house is in a rectangular form and its facade is in the northeast direction. The front facade, where the entrance of the house is also provided, overlooks the garden bordered by a 2.00-m-high wall. Other facades look outside of the garden area. House1 consists of four rooms, two sofas, a bathroom, a toilet, and storage units. The two "sofas" of the house are both disconnected and the entrances are different. In planning, geometry of both sides is in a rectangular form. Three rooms, bathrooms, kitchens, and the large "sofa" that opens to the other large "sofas" are used by all households. The big "sofa" provides a connection with the rooms and also has the function of a commonplace where the people spend their time in the summer. It is seen that households are living in rural dwellings in the region, and most of their living activities take place in a place close to the entrance of the house. Especially, this multifunctional room is actively used all day in winter months. All activities such as resting, sleeping, and eating take place in this living space. Other rooms in House1 are generally used for storage purposes, not for daily activities.

The small "sofa" and the room attached to it have been added to the plan so that they can be used by the household members who will get married. The small "sofa" does not have a feature like a big "sofa." It is a transition area that ensures that the room is connected and not opened directly to the external environment. In addition it contributes to protect the space against external conditions. With the addition of this section to the building, two different residential properties have been formed, whose entrances are planned from the same garden, planned side by side. In addition, there are toilet and storage units in the garden area. The facade and windows of the House1 are plain. There are no decorative elements of the facades or window edges. The windows that are used in the spaces are one. The window sizes (80 cm × 125 cm) of the room where daily activities take place are kept larger than the window sizes [(75 cm × 75 cm) and (55 cm × 70 cm)] of the other rooms. The "sofa" has a small window (30 cm × 50 cm) overlooking the garden. Since the walls are carriers, the window spaces are kept small. The sections and the plan of House1 structure are shown in Fig. 10. The front facade of the House1 structure is displayed in Fig. 11.

When House1 was first built, it was designed as a residence consisting of three units. The layout of the plan was created by placing the spaces at three surroundings of the middle hall. However, the need for a service space such as a kitchen and a bathroom has occurred in addition to the existing spaces over time. For this purpose, the spaces needed have been added. With the addition of these service places, the form and size of the House1 have been changed. During a meeting with the House1 users, it was stated that these changes were made in order to enlarge the building, which was made in three stages. When it

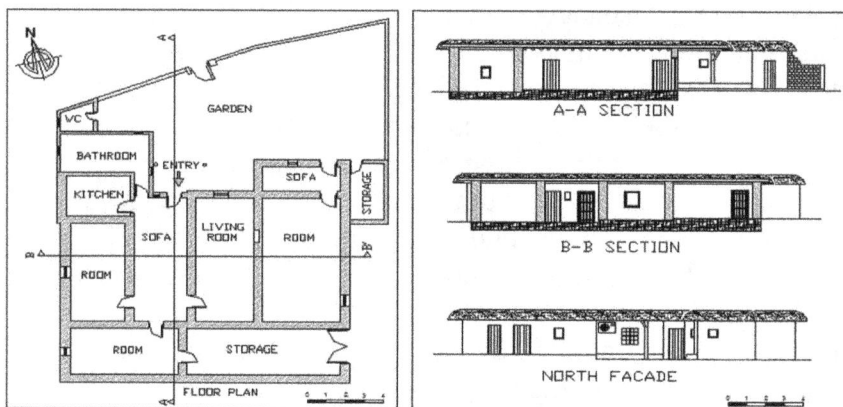

Fig. 10: The Sections and Plan of House1 Structure

Fig. 11: The Front Facade of House1 Structure

was first built, House1, which has three rooms, was 125.49 m²in area (stage 1). With the addition of a kitchen and bathroom to the existing building in stage 1, the residence has an area of 146.78 m² (stage 2).

The fact that these added spaces are located in the north, northwest directions, provided an advantage in terms of heat preservation due to the reduction of the outer wall surfaces of the room and the hall behind it. In addition, the dominant wind direction has an effect on increasing heat losses on the surfaces of the building depending on the direction of the building. When it's evaluated from this point of view, the directions in which the added spaces are located reduce the effect of heat losses by protecting the front of the house against the wind, as the dominant wind direction of the region is northwest. The number of rooms was insufficient with the expansion of the family in residential usage and the building area was enlarged to 227.04 m² (stage 3) by adding a separate sofa, room, and warehouse. The last addition made is located in the north, east, and south directions. By adding the warehouse to the back of the house, the outer surface areas of the spaces are reduced and heat losses are prevented. The changing phases of House1 structure (Phase 1, Phase 2, Phase 3) are shown in Fig. 12.

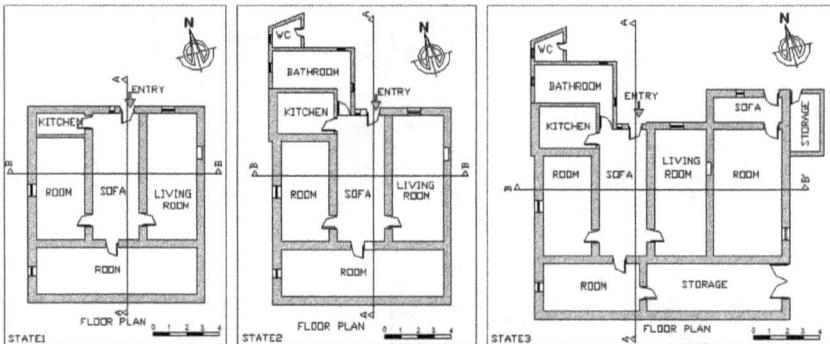

Fig. 12: The Changing Phases of House1 Structure (Phase 1, Phase 2, Phase 3)

3.1.2 *The Plan and Change Phases of House2*

The sections and plans of House2 structure are shown in Fig. 13. The entrance door of House2 opens to the garden area in the east direction. It is bordered by a 1.40-m-high garden wall. The kitchen and rooms added to the building are located along one side of the garden area in the east direction. The plan of the building was created by placing the spaces on the opposite sides of the hall. House2 consists of three rooms, a kitchen, pantry, WC, and bathroom. There are two stables and woodshed section, and WC entrance in the area behind the house. In the current plan, House2 has two "sofas." The large "sofa" is connected with two

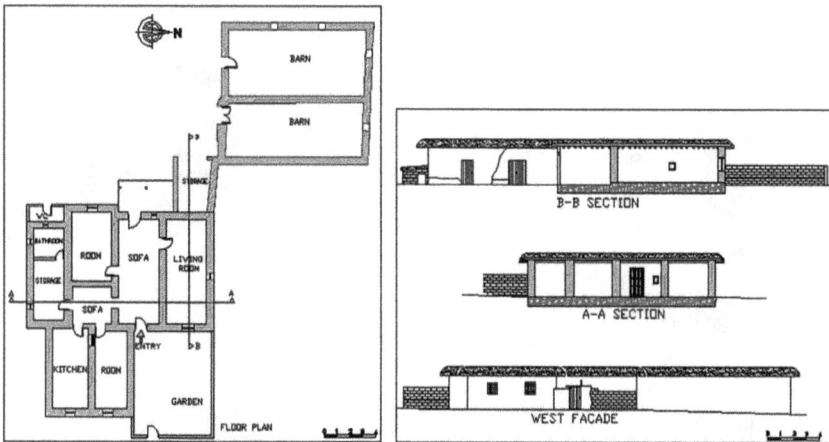

Fig. 13: The Sections and Plan of House2 Structure

rooms. The small "sofa" provides access to places such as the kitchen, room, and cellar. The small "sofa" has no relation to the external environment, but it has only a connection with the big "sofa" of House2. In the residence, which has a plan type with inner "sofas," there are door and window openings on two opposite walls of the living room facing the outer environment. With this type of planning, airflow is created in the "sofa" by keeping the doors on the opposite sides of the hall open in the summer. In the buildings of this region where the summer months are very hot, the "sofa" has a cooler microclimatic environment compared to the other rooms with this useful feature. Therefore, households spend their daily activities here in the summer. The main living room of House2 is located in on the north and east facades (Fig. 14).

There are two windows in this room. The window sizes (50 cm × 70 cm) on the north facade of the room are smaller than the window dimensions (90 cm × 90 cm) on the east facade. The small size of this window is to prevent heat losses that may occur during the cold period. A window of (75 cm × 110 cm) or (85 cm × 125 cm) is planned in other rooms of House2. The facades are plain and there is no decoration on them.

Since the House2 building was first built, it was enlarged by adding space according to the needs. The growth phase of the building has changed in four stages over time. House2 was planned built with a room where daily activities will take place

Fig. 14: Facades of House2 Structure

and a "sofa" layout that extends throughout this room (stage 1). The room is facing north, east, and west directions. The western facade of this room is covered with walls and the front side is covered with an area where wood is stored. In the second phase, there is a room with one direction and on the other side of the hall, there is a smaller "sofa" that is added in order to provide an access to rooms, warehouses, bathroom spaces, and these spaces. The added storage area was also used as a kitchen. The additions of this space were made in the south direction and the structure was shaped in a rectangular form. Windows were used on the south and west facades of the added space. In the third phase, a room and kitchen were added in the east direction of the building and access was provided for the small sofa. A window was opened to these spaces from the eastern facades. In the fourth phase, no additions were made to the mass of the house, but two large-sized spaces were built on the side of the woodshed to be used as a barn. These spaces, which are not directly related to House2 mass, prevent the wind effect coming from the northwest direction, which is the dominant wind direction in the region, from affecting the facades of the house. The positioning directions of these masses have an effect on reducing heat losses. The house that has changes in its size and form was created in four phases reaching 77.84 m² in the first phase, 145.60 m² in the second phase, 189.17 m² in the third phase, and 328.37 m² in the fourth phase (Fig. 15).

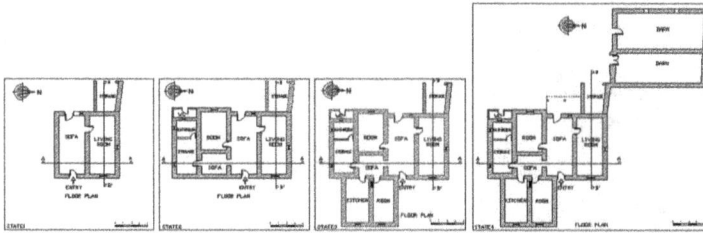

Fig. 15: Change Phases of House2 Structure (Phase 1, Phase 2, Phase 3, Phase 4)

3.1.3   The Construction Materials and Properties

The walls of the building planned as a single store are adobe and built by masonry system. The adobe used as a wall material that has heat retention and heat storage properties. The adobe provides storing of the heat in the wall during the day and discharges this heat back in the evening hours. In this way, it reduces the thermal loads of the room with the time delay effect that prevents the temperature of the zone from falling rapidly. Accordingly, the effect of the outdoor temperature in the cold period shrinks allowing it to go into the interior after a long time. The thermal capacity of the materials is directly proportional to the specific heat and the amount of mass. Therefore, the thickness of a material with high thermal capacity reinforces the effect of the thermal mass in the shell. The adobe used on the wall is approximately 45 cm. Soil plaster was used on the inner and outer wall surfaces of the building. Together with the plaster, the total wall thickness reached to 50 cm. The upper floor of the building is a flat floor created by laying approximately a 25-cm-thick mudbrick paste with straw that is added after wooden plaques or bushes, laid on wooden beams of about 12 cm diameter, are placed along the short edge of the room. The soil material used in the upper floor, which is a part of the building shell, provides time delay with its heat retention and heat storage feature, which contribute to the reduction of the amplitude of the effect of outside air temperature within a long time. In the simulation program used, the layers that make up the architectural elements and their properties such as thickness, thermal conductivity values, thermal permeability coefficients, heat capacities, and densities are shown in Table 1. Symbols and units of material are displayed in Table 2.

Table 1: Component Materials and Heat Transfer Properties

<table>
<tr><th colspan="13">THERMAL FEATURES OF MATERIALS</th></tr>
<tr><th rowspan="3">MATERIAL</th><th rowspan="3">BUILD</th><th rowspan="3">Layers Forming the Wall Section</th><th rowspan="3">d_t
m</th><th colspan="7">Thermal Properties of Wall Section Layers</th><th rowspan="3">C_p
J/kg.K</th><th rowspan="3">D
kg/m³</th></tr>
<tr><th colspan="2">Layer-1</th><th colspan="2">Layer-2</th><th colspan="2">Layer-3</th><th rowspan="2">U
W/m² K</th></tr>
<tr><th>d
m</th><th>λ_h
W/m K</th><th>d
m</th><th>λ_h
W/m K</th><th>d
m</th><th>λ_h
W/m K</th></tr>
<tr><td colspan="13">WALL</td></tr>
<tr><td>ADOBE</td><td>House1</td><td>0.02-m Soil Plaster (exterior)</td><td rowspan="3">0.49</td><td>0.02</td><td>0.7</td><td>0.5</td><td>0.75</td><td>0.02</td><td>0.7</td><td rowspan="3">1.2</td><td>840</td><td>1600</td></tr>
<tr><td></td><td></td><td>0.45-m Adobe Wall</td><td></td><td></td><td></td><td></td><td></td><td></td><td>880</td><td>1730</td></tr>
<tr><td></td><td></td><td>0.02-m Soil Plaster (interior)</td><td></td><td></td><td></td><td></td><td></td><td></td><td>840</td><td>1600</td></tr>
<tr><td></td><td>House2</td><td>0.02-m Soil Plaster (exterior)</td><td rowspan="3">0.59</td><td>0.02</td><td>0.7</td><td>0.6</td><td>0.75</td><td>0.02</td><td>0.7</td><td rowspan="3">1.03</td><td>840</td><td>1600</td></tr>
<tr><td></td><td></td><td>0.55-m Adobe Wall</td><td></td><td></td><td></td><td></td><td></td><td></td><td>880</td><td>1730</td></tr>
<tr><td></td><td></td><td>0.02-m Soil Plaster (interior)</td><td></td><td></td><td></td><td></td><td></td><td></td><td>840</td><td>1600</td></tr>
</table>

THERMAL FEATURES OF MATERIALS

MATERIAL	BUILD	Layers Forming the Wall Section	d_t m	Thermal Properties of Wall Section Layers							C_p J/kg.K	D kg/m³
				Layer-1		Layer-2		Layer-3				
				d m	λ_h W/m K	d m	λ_h W/m K	d m	λ_h W/m K	U W/m² K		
UPPER FLOOR												
SOIL ROOF	House1	0.25-m soil	0.371	0.25	0.837	0	0.14	0.12	0.16	0.77	1046	1300
	House2	0.01-m wood plaque									1890	700
		0.12-m wood beam									1260	720
FLOORING												
CONCRETE	House1-House2	0.10-m lean concrete	0.2	0.1	1.4	0.2	1.3	-	-	2.74	650	2100
		0.15-m blockade									920	2240

Table 2: Symbols and Units of Material Thermal Properties

Explanation	Symbol	Unit
Total wall thickness	d_t	m
Thickness of the building component	d	m
Thermal conductivity calculation value	λ_h	W/m.K
Total thermal permeability coefficient of the building component	U	W/m².K
Heat capacity	C_p	J/kg.K
Density	D	kg/m³

3.2   *Simulation Program and Parameters*

Within the scope of the study, the energy performance evaluation of the building was made with Ecotect v5-20 simulation program. Ecotect v5-20 was developed by Andrew Marsh at the University of Western Australia's Architectural Science Laboratories. This program is designed to measure energy performance in the early design phases [20,21]. It enables the evaluation of the heating, air conditioning, and ventilation systems in the building with environmental data. Ecotect v5-20 is a building performance simulation program that can perform calculations on an hourly basis and for each location in addition to its ability to analyze the performance of the building as an integrated whole. The Ecotect v5-20 program uses the "Admittance Method" that was defined in CIBSE Guide (CIBSE, 1988) for thermal performance calculations [22,23]. While this method is based on continuous state analysis, it could be used as an admittance, surface factor, and amplitude reduction factor, and the U-value to determine the thermal performance to enable the building's dynamic response to be simulated [24,25]. First, the materials used in the building and their properties are defined in the simulation program. In the energy analysis, it is assumed that the household using the structure consisted of four people.

In the structures examined, there are places that have function in line with the social life of the local people and have separate functions. These spaces are used in living space, rooms used for storage purposes, sofa, kitchen, bathroom, and stable functions. In the analysis, data entries were made according to the characteristics of the households using the places. The local people meet all the vital activities such as resting, eating, and sleeping in the living space. Most of the daily time frame of household users, especially in the cold periods of the year, takes place in this place. Therefore, the need for heating and cooling with an active system is required only in the living space and there is no need for heating–cooling system in the other units in the building. For this reason, each unit forming the structure was created as a separate "zone" in the energy analysis. Active heating and cooling have been defined in the living spaces of the buildings and other spaces have been evaluated according to their natural climate conditions. The air exchange rate in the spaces has been entered as 0.7 n_h according to TS 825. It is assumed that the doors and windows are closed during the cold months of the year, opened between 20:00 and 08:00 in the hot months of the year, and closed at other times. With this acceptance, it is ensured that the outside air, which is low at night, is stored indoors in the warm months of the year and the effect of the high outdoor temperature during the day is reduced.

In heating and cooling load analysis, the comfort conditions range was accepted as 18°C–26°C. It is assumed that the active system used is effective between 08:00 and 22:00 during the day and the active energy usage will be activated at temperatures outside the comfort conditions. Since the usage of the sofa is active in the hot summer period, it is defined by adding the functional features of the activity inputs in the summer months. For service units such as bathrooms and kitchens, data have been entered by determining their usage according to their functional features. House1 and House2 structures according to Ecotech software are shown in Fig. 16.

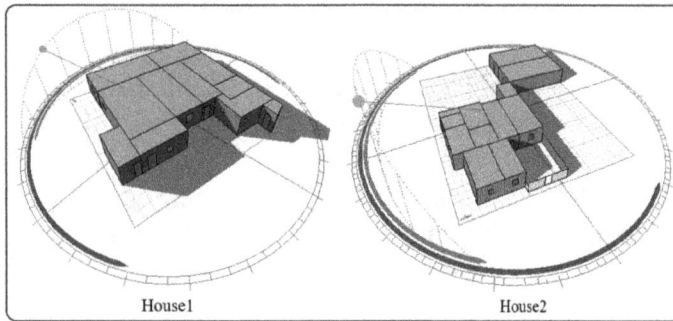

Fig. 16: House1 and House2 Structures According to Ecotech Software [22]

4.0   Results and Discussions

4.1   *Energy Performance Analysis of House1 in Summer and Winter Months*

In Tables 3 and 4, the growth phases of House1 consisting of three stages and the internal temperature values of the hottest and coldest days of the residence according to these phases are given. It was observed that the internal temperature value did not change on the hottest day of the year in the living space of the building, which changed its shape with its growth stages. On the coldest day of the year, it is seen that the internal temperature value of the living space during the day did not change in the first and second phases, but it changed in the third phase. Since no unit was added around the living space in the first and second phases of the building, there was no change in the outer surface area. In the third phase, the exterior wall surface area of the room, which was added next to the long wall of this rectangular space associated with the external environment, was reduced. As the surface outer areas of the buildings become smaller, the amount of the heat that lost from the building shell decreases and provides an advantage in terms of heat preservation. It is seen that this formal change contributes to the reduction of the need for heating during the coldest period of the year. The external temperature values of the House1 living space (the coldest and hottest days of the year) are displayed in Fig. 17.

Table 3: The Internal Temperature Values of the House1 Living Area (the Hottest Day of the Year)

	Stage 1	Stage 2	Stage 3	
Stages of Growth				
Hour	Internal	Internal	Internal	Outside
	temperature °C	temperature °C	temperature °C	temperature °C
0	26.7	26.7	26.7	28
1	25.4	25.5	25.5	27.1
2	24.8	24.8	24.8	26.6
3	24.3	24.3	24.3	26.2
4	24.5	24.5	24.6	26.4
5	27.2	27.2	27.2	28.4
6	30.4	30.4	30.4	30.8
7	33.8	33.8	33.8	33.4
8	33.9	33.9	33.9	35.9
9	33.9	33.9	33.9	38.2
10	33.8	33.8	33.8	40.1
11	33.8	33.8	33.8	41.5
12	33.8	33.8	33.8	42.6
13	33.8	33.8	33.8	43.1
14	33.8	33.8	33.8	43.2
15	33.8	33.8	33.8	42.6
16	33.8	33.8	33.8	41.5
17	33.8	33.8	33.8	39.9

(Continued)

Table 3: The Internal Temperature Values of the House1 Living Area (the Hottest Day of the Year) (*Continued*)

	Stage 1	Stage 2	Stage 3	
18	33.7	33.7	33.7	38.1
19	33.8	33.8	33.8	36.4
20	33.8	33.8	33.8	34.7
21	33.2	33.2	33.2	33
22	30.9	30.9	30.9	31.2
23	28.6	28.6	28.6	29.5

Table 4: The Internal Temperature Values of the House1 Living Area (the Coldest Day of the Year)

	Stage 1	Stage 2	Stage 3	
Stages Of Growth				
Hour	Internal	Internal	Internal	Outside
	temperature °C	temperature °C	temperature °C	temperature °C
0	0.7	0.7	1.6	−4.4
1	0.6	0.6	1.5	−5.2
2	0.5	0.5	1.5	−5.9
3	0.5	0.5	1.4	−6.3
4	0.4	0.4	1.4	−6.7
5	0.4	0.4	1.3	−7
6	0.4	0.4	1.3	−7.1
7	0.4	0.4	1.4	−6.3
8	0.6	0.6	1.5	−5.1

	Stage 1	Stage 2	Stage 3	
9	0.7	0.7	1.7	−4
10	0.7	0.7	1.7	−3.7
11	0.7	0.7	1.7	−3.4
12	0.7	0.7	1.7	−3.4
13	0.7	0.7	1.6	−3.9
14	0.6	0.6	1.6	−4.5
15	0.5	0.5	1.5	−5.1
16	0.5	0.5	1.4	−5.6
17	0.4	0.4	1.4	−5.9
18	0.4	0.4	1.4	−6.1
19	0.4	0.4	1.3	−6.4
20	0.3	0.3	1.3	−6.6
21	0.3	0.3	1.3	−6.9
22	0.3	0.3	1.3	−7.1
23	0.3	0.3	1.3	−7.3

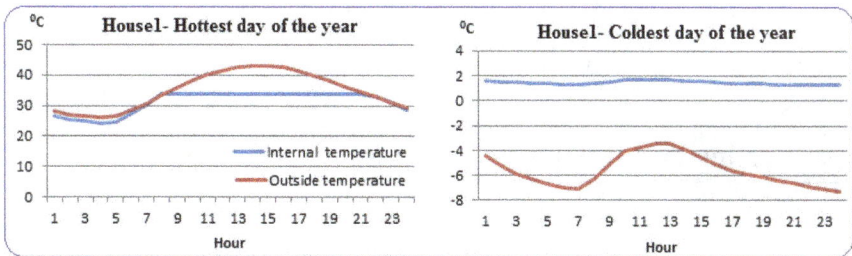

Fig. 17: The External Temperature Values of the House1 Living Space (the Coldest and Hottest Days of the Year)

Table 5 shows quantitative information about the physical changes of the House1 according to the growth stages. The area of the building form was changed as 125.49 m² in the first phase, 146.78 m² in the second phase, and 227.04 m² in the third phase. While the building form was evaluated, the "A/V ratio" was

Table 5: Growth Stages of House1 Structure

Stages of Growth	Stage 1	Stage 2	Stage 3
Housing Data			
Building area (m²)	125.49	146.78	227.04
Volume (m³)	326.27	381.63	590.30
Total outer wall surface area (m²)	114.89	143.30	192.37
Total outer surface area (m²)	240.38	290.08	419.41
Total surface area (m²)	509.75	640.60	984.46
Area/volume (total outer surface Aa	0.74	0.76	0.71
Area/volume (outer wall surface area) (m²)	0.35	0.38	0.33
Area of the opaque component (m²)	113.56	139.82	187.98
Transparent component area (m²)	1.33	3.48	4.39
Heating load (kWh/m²)	7.42	6.40	4.09
Cooling load (kWh/m²)	7.14	6.16	4.22
Total energy (kWh/m²)	14.56	12.56	8.31

calculated and compared with the total annual energy amount. The "A/V ratio" is defined as the ratio of the total heat loss area (A) of the building form to the volume (V) of the building protected against heat losses. According to the results of the analysis, depending on the changed form of the building, the annual total energy amount decreases with the decrease of the "outer surface A/V ratio" of the building. The annual total energy amount decreases with the decrease of the "outer surface A/V" ratio depending on the changed form of the building according to the results of the analysis. The A/V ratio has diminished from 0.74 to 0.71, and the annual total energy requirement, which was 14.56 kWh/m² in the first phase, has decreased to 8.30 kWh/m² in the third phase (Fig. 18). The annual heating energy of the building has decreased from 7.42 kWh/m² to 4.09 kWh/m², and the annual cooling energy has decreased from 7.14 kWh/m² to 4.22 kWh/m². The annual energy requirement of House1 is decreased by 13.74% from the first phase to the second phase and by 33.84% from the second phase to the third phase. When the first phase and the third phase of the building are compared, the annual total needed energy is decreased by 42.93%.

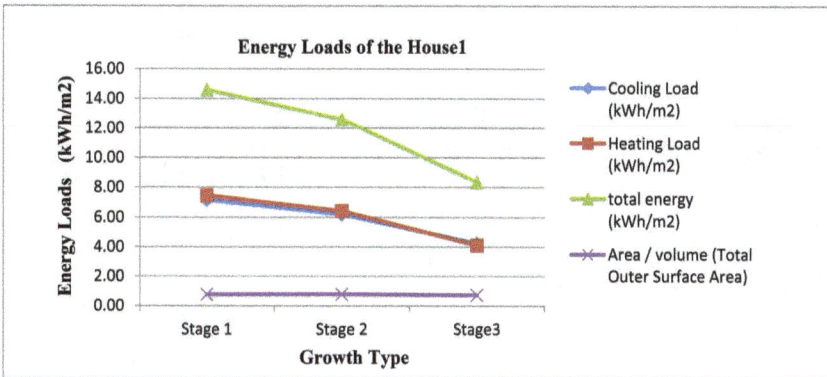

Fig. 18: Relationship Between the Growth Phase and the Energy Loads of the House1 Structure

4.2   *Energy Performance Analysis of House2 in Summer and Winter Months*

In Tables 6 and 7, the growth phases of House2 and the temperature values of building an internal place, according to the hottest and the coldest day of the year, are given. The building form has changed by adding new space in each phase. The additions made to develop the horizontal and vertical axis arrangement in the south and west directions of the building. Since this development did not change the outer wall area of the living space, it did not cause a difference in temperature changes in this space (Fig. 19). Depending on the thermophysical feature

Table 6: The Internal Temperature Values of the House2 Living Area (the Hottest Day of the Year)

Stages of Growth	Stage 1	Stage 2	Stage 3	Stage 4	
Hour	Internal Temperature °C	Internal Temperature °C	Internal Temperature °C	Internal Temperature °C	Outside Temperature °C
0	25.5	25.5	25.5	25.5	28
1	23.9	23.9	23.9	23.9	27.1
2	23.2	23.2	23.2	23.2	26.6
3	22.5	22.5	22.5	22.5	26.2
4	22.8	22.8	22.8	22.8	26.4
5	25.9	25.9	25.9	25.9	28.4
6	29.6	29.6	29.6	29.6	30.8
7	33.6	33.6	33.6	33.6	33.4

	Stage 1	Stage 2	Stage 3	Stage 4	
8	33.7	33.7	33.7	33.7	35.9
9	33.8	33.8	33.8	33.8	38.2
10	33.6	33.6	33.6	33.6	40.1
11	33.6	33.6	33.6	33.6	41.5
12	33.7	33.7	33.7	33.7	42.6
13	33.7	33.7	33.7	33.7	43.1
14	33.8	33.8	33.8	33.8	43.2
15	33.8	33.8	33.8	33.8	42.6
16	33.8	33.8	33.8	33.8	41.5
17	33.9	33.9	33.9	33.9	39.9
18	33.8	33.8	33.8	33.8	38.1
19	33.9	33.9	33.9	33.9	36.4
20	33.9	33.9	33.9	33.9	34.7
21	33.2	33.2	33.2	33.2	33
22	30.4	30.4	30.4	30.4	31.2
23	27.7	27.7	27.7	27.7	29.5

Table 7: The Internal Temperature Values of the House2 Living Area (the Coldest Day of the Year)

Stages of growth	Stage 1	Stage 2	Stage 3	Stage 4	
	Internal	Internal	Internal	Internal	Outside
Hour	Temperature °C	Temperature °C	Temperature °C	Temperature °C	Temperature °C
0	-1.2	-1.2	-1.2	-1.2	-4.4
1	-1.3	-1.3	-1.3	-1.3	-5.2
2	-1.5	-1.5	-1.5	-1.5	-5.9
3	-1.6	-1.6	-1.6	-1.6	-6.3
4	-1.7	-1.7	-1.7	-1.7	-6.7
5	-1.7	-1.7	-1.7	-1.7	-7
6	-1.8	-1.8	-1.8	-1.8	-7.1
7	-1.7	-1.7	-1.7	-1.7	-6.3
8	-1.5	-1.5	-1.5	-1.5	-5.1

	Stage 1	Stage 2	Stage 3	Stage 4	
9	-1.4	-1.4	-1.4	-1.4	-4
10	-1.4	-1.4	-1.4	-1.4	-3.7
11	-1.4	-1.4	-1.4	-1.4	-3.4
12	-1.4	-1.4	-1.4	-1.4	-3.4
13	-1.5	-1.5	-1.5	-1.5	-3.9
14	-1.5	-1.5	-1.5	-1.5	-4.5
15	-1.6	-1.6	-1.6	-1.6	-5.1
16	-1.5	-1.5	-1.5	-1.5	-5.6
17	-1.6	-1.6	-1.6	-1.6	-5.9
18	-1.6	-1.6	-1.6	-1.6	-6.1
19	-1.7	-1.7	-1.7	-1.7	-6.4
20	-1.7	-1.7	-1.7	-1.7	-6.6
21	-1.8	-1.8	-1.8	-1.8	-6.9
22	-1.9	-1.9	-1.9	-1.9	-7.1
23	-1.9	-1.9	-1.9	-1.9	-7.3

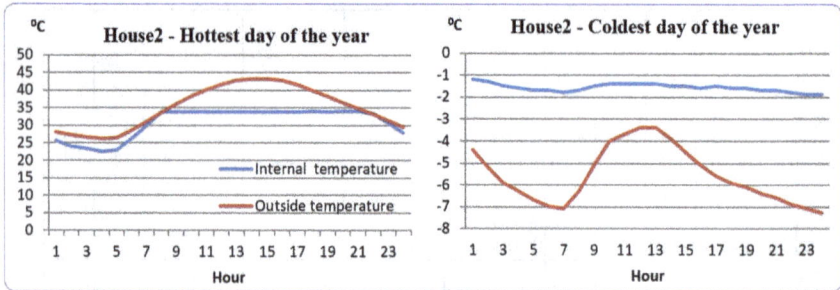

Fig. 19: The External Temperature Values of the House2 Living Space (the Coldest and Hottest Days of the Year)

of the adobe, which is the wall material, the temperature stored in the wall shows that more stable temperature changes occur without any rapid indoor temperature changes during the day with the effect of time delay and amplitude reduction (Tables 6 and 7).

The structural growth stages of House2 can be seen in Table 8. The area of the House2, whose form was changed in four stages, changed to 77.84 m² in the first phase, 145.60 m² in the second phase, 189.17 m² in the third phase, and 328.37 m² in the fourth phase. While calculating the A/V ratio in the analysis, the stables are not included in the A/V ratio calculation since the barn units added in the fourth stage do not change the outer surface area of the residential structure. Since the barn units added in the fourth phase do not change the outer surface area of the residential structure, the barns are not included in the A/V ratio calculation. It is seen that the total amount of energy decreases with the decrease in the "outer surface area/volume" ratio depending on the House2 form change according to the results of the analysis. The A/V ratio fell from 1.23 to 1.09, and the annual total energy requirement, which was 64.15 kWh/m² in the first phase, decreased to 15.42 kWh/m² in the fourth phase (Fig. 20). The annual heating energy of the building has decreased from 34.69 kWh/m² to 8.25 kWh/m² and the annual cooling energy has decreased from 29.46 kWh/m² to 7.17 kWh/m². The location of the spaces added to the building in the south and east directions has increased the average radiation temperature by affecting the amount of solar radiation, which affects the surface of the building. This can be interpreted as a result of the heating

Table 8: Growth Stages of House2 Structure

Stages of Growth	Stage 1	Stage 2	Stage 3	Stage 4
Housing Data				
Building area (m²)	77.84	145.60	189.17	328.37
Volume (m³)	210.17	393.12	510.76	886.60
Total outer wall surface area (m²)	180.88	290.36	368.01	634.65
Total outer surface area (m²)	258.72	435.96	557.18	963.02
Total surface area (m²)	321.90	655.38	847.92	1320.16
Area/volume (total outer surface area, m²)	1.23	1.11	1.09	1.09
Area/volume (outer wall surface area, 1/m)	0.86	0.74	0.72	0.72
Area of the opaque component (m²)	179.53	287.80	364.13	630.77
Transparent component area (m²)	1.35	2.56	3.88	3.88
Heating load (kWh/m²)	34.69	18.58	14.32	8.25
Cooling load (kWh/m²)	29.46	15.94	12.34	7.17
Total energy (kWh/m²)	64.15	34.52	26.66	15.42

load being reduced more than the cooling load. The annual energy requirement of House2 is decreased by 46.19% from the first phase to the second phase, by 22.77% from the second phase to the third phase, and by 42.16% from third phase to fourth phase. When the first phase of the building and the fourth phase are compared, the annual total energy need is decreased by 75.46%.

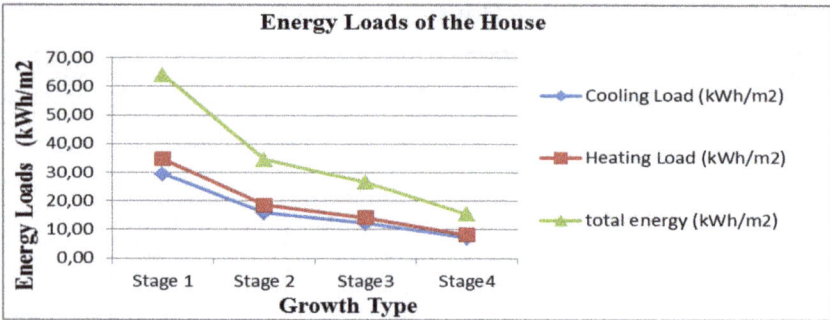

Fig. 20: Relationship Between the Growth Phase and the Energy Loads of the House2 Structure

5.0   Conclusions

It is important to design the building form in accordance with the climate characteristics of the region where the building is located in order to reduce energy costs. Rural houses have planning features compatible with the climate in this respect. In addition, in rural settlements, the building is enlarged by adding new spaces of most of the buildings in line with the needs, and the form of the structure changes as well. Changing the forms of the structures, and correspondingly increasing or decreasing the outer surface area, are effective in the energy performance of the structure. For this purpose, two residential buildings built with traditional materials in rural settlements were divided into phases in terms of form change.

In the analysis made for House1 and House2, depending on the change of the outer wall surface area of the living space where the households spend almost all their activities the temperature values of the coldest and hottest days of the year, were examined. Furthermore, it It was observed that the temperature of the living space in House1 increased with the decrease in the outer wall surface area and the internal temperature values increased on the coldest day of the year. On the hottest day of the year, it was observed that the internal temperature value of this place was significantly lower than the external temperature value. In addition, since the additions made in House2 do not reduce the outer wall surface of the living space, there was no difference in the internal temperature values during the cold and hot days of the year. In order to contribute to the reduction of heat losses in the cold periods of the year and to provide comfort conditions in the interior, it is recommended to develop designs that reduce the outer surface area of the buildings.

In addition, performances can be compared thanks to the analysis results obtained by taking into account the annual heating, cooling and total energy amount, which changes depending on the change of building form. In simulation analysis, it has been observed that the decrease in the outer surface A/V ratio reduces the total annual needed energy. In order to contribute to minimizing energy consumption, it is recommended to consider the A/V ratio in the need to enlarge the building by adding units or in new designs. While the annual energy need of House1 was 14.56 kWh/m^2 in the first phase, it decreased to 831 kWh/m^2 in the third phase. While the annual energy requirement of House2 was 64.15 kWh/m^2 in the first phase, it decreased to 15.42 kWh/m^2 in the fourth phase.

As a consequence, the analysis focused on the architecture criteria of the Diyarbakır rural houses reveals that building form and settlement structure affect energy efficiency to ensure optimum conditions. The simplified schematic built to measure the results of the variables selected shows that the correct construction parameter values guide the building toward an energy-efficient configuration. In line with these findings, it is openly shown that the planning process should begin with the study of environmental factors and that the building should take form according to climate information. Energy-efficient technologies taken from conventional houses into modern building practices may be viewed as a groundbreaking step in developing potential energy-efficient contemporary settlements.

References

(1) ASHRAE Standard 55-2004. Thermal Environmental Conditions for Human Occupancy. ASHRAE Standard. 2004.

(2) Burberry P. *Building for Energy Conservation*. Michigan University, MI: Architectural Press; 1978:59. ISBN0470993502,9780470993507

(3) Balo F, Polat H. Green design effectiveness for a mini automotive-repair facility. In: Stagner JA, Ting DS-K, eds. *Green Energy and Infrastructure: Securing a Sustainable Future*. Boca Raton, FL: Taylor & Francis Group (CRC Press); 2020.

(4) Balo F. Feasibility study of "green" insulation materials including tall oil: environmental, economical and thermal properties. *Energy Build*. 2015;86:161–175.

(5) Balo F. Energy and economic analyses of insulated exterior walls for four different city in Turkey. *Energy Edu Sci Technol Part A: Energy Sci Res*. 2011;26(2):175–188.

(6) Balo F. Castor oil-based building materials reinforced with fly ash, clay, expanded perlite and pumice powder. *Ceram Silik*. 2011;55(3):280–293.

(7) Balo F. Characterization of green building materials manufactured from canola oil and natural zeolite. *J Mater Cycle Waste Manag*. 2015;17:336–349.

(8) Hensen JLM. *On the Thermal Interaction of Building Structure and Heating and Ventilating System* [PhD Dissertation]. Glasgow, Scotland: Energy Systems Research Unit, Department of Mechanical Engineering, University of Strathclyde; 1991.

(9) Balo F, Polat H. *Energy Productivity With the Proper Medical Waste Storage Design*. Irvine, CA: Brown Walker Press; 2020.

(10) Doğan E, Polat H. A research for efficiency of using prefabrication building components in Building Information Modeling BIM process. *Int Multiling Acad J*. 2016;3(4):1–7.

(11) Ulukavak HG. *Enerji Performansı Öncelikli Mimari Tasarım Sürecinin İlk Aşamasında Kullanılabilecek Tasarıma Destek Değerlendirme Modeli* [Doktora Tezi]. Ankara, Turkey: Gazi Üniversitesi, Fen Bilimleri Enstitüsü; 2009.

(12) Manioğlu G, Oral KG. Effect of courtyard shape factor on heating and cooling energy loads in hot-dry climatic zone. *Energy Procedia*. 2015;78:2100–2105.

(13) Muhaisen AS, Gadi M. Effect of courtyard proportions on solar heat gain and energy requirement in the temperate climate of Rome. *Build Environ*. 2006;41:245–253.

(14) Darkot B, Arslan R. *The History of Diyarbakır City and Its Present Position*. Turkey: Diyarbakır Culture Guide; 2000: 48.

(15) Beysanoğlu Ş. *History of Diyarbakır with its Monuments and Inscriptions I*. Ankara, Turkey: Yapı Kredi Publishing; 1996.

(16) Ergin Ş, Polat H. Material use in Diyarbakir rural architecture. *Kerpic'19 – Earthen Heritage, New Technology, Management, 7th*

International Conference, September 5–7, 2019, Köyceğiz - Muğla, Turkey. http://www.kerpic.org/2019/venue.html

(17)   TS 825. *Thermal Insulation Rules in Buildings*. Ankara, Turkey: TürkishStandartları Institute; 2013.

(18)   https://www.mgm.gov.tr. Accessed February 1, 2020.

(19)   http://www.mgm.gov.tr/iklim. Accessed February 20, 2020.

(20)   https://1.bp.blogspot.com/-KQEI5JDC8U/Vv2OJS390LI/ AAAAAAAAeTY/92rcGK7G62UZ99iSpGnr_Uv7PYbwxbbKA/ s1600/diyarbakir_ili_ilceleri.jpg. Accessed February 15, 2020.

(21)   http://www.squ1.com. Accessed February 20, 2020.

(22)   Çengel YA. *Heat Transfer, A Practical Approach*. New York, NY: WCB/McGraw-Hill; 1998.

(23)   https://www.cibse.org/knowledge/knowledge-items/ detail?id=a0q20000008Jr5RAAS.Accessed February 12, 2020

(24)   Muhaisen AS. Shading simulation of the courtyard form in different climatic regions. *Build Environ*. 2006;41:1731–1741.

(25)   https://autodesk-ecotect-analysis.software.informer.com/1.0/. Accessed February 28, 2020.

CHAPTER 7
Sustaining Our Air and Water

Peter Brimblecombe
[1]*Department of Marine Environment and Engineering, National Sun Yat-sen University, Kaohsiung, Taiwan*
[2]*Aerosol Science Research Center, National Sun Yat-sen University, Kaohsiung, Taiwan*

Abstract

Pollution is a global problem, yet represented by numerous smaller issues at a local level. Greenhouse warming is a global issue that despite its increasing impact remains debated. Regionally, acid rain damaged the forests and lakes of Europe and North America, but was successfully addressed by controlling emissions. Nevertheless, it persists and has become characteristic of China and India. The upper parts of the atmosphere are contaminated by chlorine derived from refrigerants that enhance ozone depletion, though international agreements have reduced this problem. Biomass burning, volcanoes, and windblown dust are seemingly natural processes, yet cause widespread health problems and disrupt air traffic. In the oceans, oil pollution has long been a major problem, although in the current decade it is plastic pollution that has come to dominate public concern. Local air pollution problems typify cities, but are also found around large industrial plants. Air pollutants arise directly as exhaust gases, but are also formed from reactions in the atmosphere, which lead to photochemical smog. Cities additionally suffer from urban runoff that runs across hard surfaces, such as roads and leads to flooding and polluted water. Factories, sewage works, and large point sources add to water pollutants. Legal and fiscal responses, add to technical controls as potential solutions to environmental problems.

Keywords: Air, water, acid rain, ozone layer

1.0 Introduction

Few are unaware of the contemporary threats that face our air and water. These seem to feature so regularly in the news media that it is easy to either ignore them or be overwhelmed by a sense of crisis. It seems at times that nowhere on the globe is there an environment that is unaffected. Pesticides are found in the Arctic, plastic debris on remote Pacific atolls, our refrigerants in the stratosphere, and carcinogens in coastal sediments. There appear to be insurmountable threats to our world, yet this book aims to explore how tomorrow can be sustained through understanding and innovation.

Clearly describing global environmental pressures to air and water is a task that would readily stray well beyond a single chapter. The focus here will have to be aimed at broader concepts and sometimes specific details and examples might seem slight, but the accompanying references should help explore issues in greater depth.

2.0 Scale of the Problem

Pollution has to be considered in terms of its spatial scale. At the global level, it is recognized almost immediately that it is necessary to split any discussion between two enormous reservoirs, the atmosphere and the oceans; both at risk as a result of polluted air and water. These reservoirs are both fluids, with the oceans at 1.4×10^{24} g (1.4 quintillion ton) more massive than the atmosphere at 0.0052×10^{24} g. Pollutants that enter these two giant reservoirs are diluted because of their enormous size, but because the atmosphere is so much smaller than the oceans, a given amount of pollutant would be potentially diluted to a greater extent in the oceans. This could mean that the atmosphere would be more polluted than the oceans. However, both these reservoirs are fluids, so they mix relatively easily, but the mixing time for the oceans is slower, some 1,600 years, compared with the lower atmosphere, which mixes in a matter of days. Thus a temporal scale has been added to the spatial one. The rapid mixing of the atmosphere can be illustrated by the radioactive material from the Chernobyl accident in April 1986, which spread around the globe and reached the US West Coast after 10 days.

Pollutants often have characteristic times, or residence times that describe how long they are likely to persist in the atmosphere or the oceans. Most pollutants are removed over time, some rapidly and some slowly; just a few appear as though they are not removed at all. These are the most dangerous as they accumulate, thus the metal lead has accumulated in roadside soils

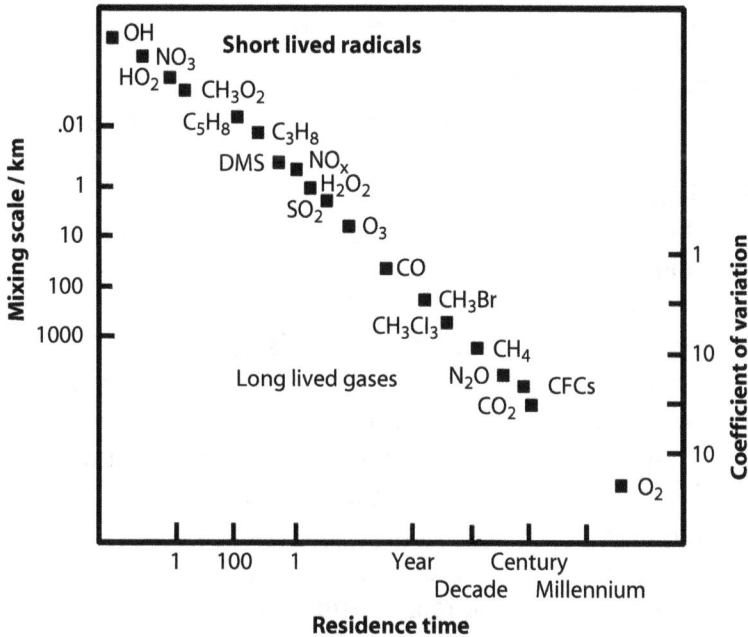

Fig. 1: The Relationship Between Lifetimes of Atmospheric Gases and Their Spatial Variability. This Can Be Considered in Terms of Physical Distance or as the Coefficient of Variation (Standard Deviation/Mean) of Concentrations.

because it is diluted only slowly, while the chlorofluorocarbons (CFCs or freons) are removed so slowly from the lower atmosphere that they accumulate in the stratosphere. Although, the residence (average lifetime) and mixing times (rate of dilution) are slightly different parameters, they are found linked in a concept known as Junge's spatial variability rule. The longer the residence time the less spatial variability found in the pollutant concentrations. This is shown for the atmosphere in Fig. 1, but can equally be applied to the oceans or other major reservoirs.

3.0 Global and Regional Pollution

At a global or regional scale, there are plenty of examples of pollution, so we can only give a selection here. These relate to some of the most important contemporary issues.

3.1. Carbon Dioxide and Climate Change

The problem of increasing concentrations of greenhouse gases and particles in the atmosphere has long been recognized along with its potential to change our climate. Fossil fuel combustion, which releases carbon dioxide into the atmosphere, has caused observable changes in its concentrations over recent centuries. This accelerated in the late 20th century such that concentrations are about 415 ppm compared to those of the 19th century, where it was close to 295 ppm. Carbon dioxide is so abundant and has the greatest capacity to absorb outgoing infrared radiation (longwave), so it forms a kind of blanket that retains heat and causes the world to warm; the radiative forcing amounts to just over 2 W m^{-2}, which leads to the recent decade being more than 0.9°C warmer than the 19th century.

The concentrations as shown in Fig. 2 along with a line of a similar form showing the temperature increase. The issue is a little complex because of numerous absorption bands in the spectra of many gases, but nevertheless there is a convincing agreement between the rise in temperature and the concentration of carbon dioxide. However, carbon dioxide is not the only gas that absorbs outgoing radiation, and some of the others are more

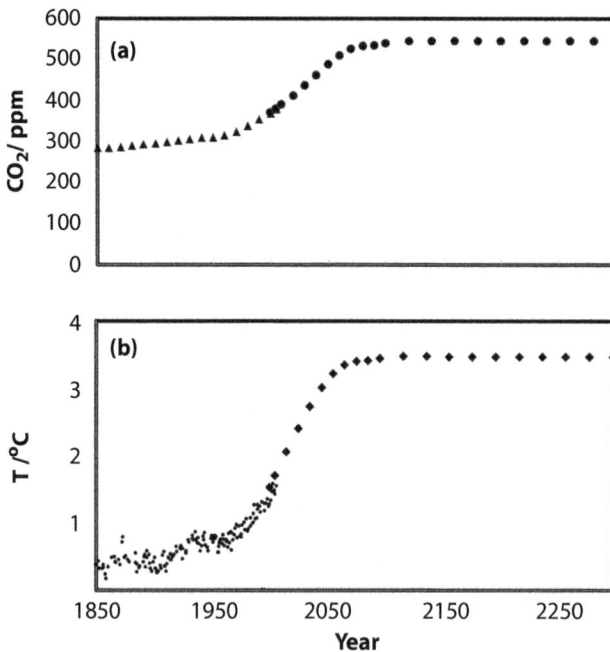

Fig. 2: (a) The Projected Concentration of CO_2 Under Scenario RCP 4.5 and (b) Observed and Projected Changes in Temperature 1850–2300.

powerful absorbers, so although their concentrations are lower they still have substantial influence. Water vapor is also a greenhouse gas, but as much of water vapor is from evaporation, it is less directly a function of human activities.

Methane is an active greenhouse gas. Its best known sources being wetlands, rice paddies, animals (from cows to termites), and melting permafrost. Additionally, there is some leakage from coal mining and natural gas use. The flux of methane to the atmosphere over time has been increasing, but it is a little more variable than carbon dioxide.

Less well known is nitrous oxide (N_2O), which has natural sources such as thawing permafrost, but increasingly agriculture and fertilized fields also represent a source. Here proteins can break down in the denitrification process producing nitrous oxide. Modern catalytic converters on automobiles can sometimes lead to increased emissions. Ozone is also a greenhouse gas and, along with the CFCs and hydrofluorocarbons (including HCFCs and

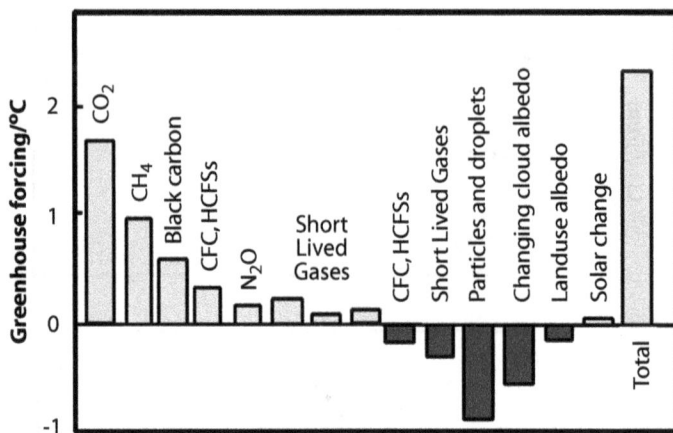

Fig. 3: Radiative Forcing by Greenhouse Gases, Particles and Albedo (Reflectance)

HFCs), contributes to global warming. Their relative contributions to the greenhouse effect are shown in Fig. 3.

3.2. Acid Rain, Snow, and Lakes

In the mid-19th century Norwegians became aware that pollution from an industrial Britain was evident as deposits on snow fields in the remote mountains. Observations of damage to fish stocks and later forests took longer to emerge, but chemical analysis of the constituents of rainfall showed that pollutants from coal burning were increasingly affecting its composition, which made it more acidic. The idea of acid rain emerged as a global issue at the *United Nations Conference on the Human Environment*, in Stockholm in 1972. Subsequent decades saw increasing concern and the development of the *Convention on Long-Range Transboundary Air Pollution* of 1979.

The initial problem largely arose from the presence of sulfuric acid in precipitation. This was caused by the high sulfur content of the coal burnt across Europe and North America. When coal is burnt the sulfur present as a solid (s), this impurity in coal is oxidized to sulfur dioxide and appears in the exhaust gas (g):

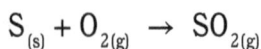

$$S_{(s)} + O_{2(g)} \rightarrow SO_{2(g)}$$

while SO_2 is only slightly soluble some dissolves in cloud drops:

$$H_2O_{(l)} + SO_{2(g)} \leftrightarrow H^+_{(aq)} + HSO^-_{3(aq)}$$

and the resultant bisulfite (HSO 3 –) in the aqueous droplets (aq) subsequently oxidizes to the stronger sulfuric acid, a reaction catalyzed by metals or directly oxidized by dissolved hydrogen peroxide (H_2O_2):

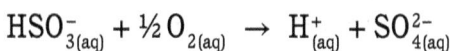

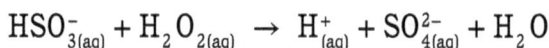

$$HSO^-_{3(aq)} + \tfrac{1}{2}O_{2(aq)} \rightarrow H^+_{(aq)} + SO^{2-}_{4(aq)}$$

$$HSO^-_{3(aq)} + H_2O_{2(aq)} \rightarrow H^+_{(aq)} + SO^{2-}_{4(aq)} + H_2O$$

It is this transformation that effectively results in acid rain.

It is worth considering that these changes represent an amplification process that can result in a significant impact of this pollution, often thousands of kilometers from its source. Initially, as the sulfur dioxide has dissolved from the air there is effectively a huge decrease in the volume it occupies. There is typically less than a cubic centimeter of water in every cubic meter of air. This is a phase volume ratio of 1:1,000,000 and thus as the sulfur is transferred from the gas to liquid phase it represents a million-fold increase in concentration; further the weak sulfurous acid (represented as bisulfite in reactions above) is transformed to the much stronger sulfurous acid, strongly enhancing the acidity of the droplets.

In the winter, some of the precipitation falls as snow. As snow crystals form from freezing water, the solutes such as sulfuric acid are concentrated in the liquid that remains as ice forms. Ultimately the acids and other pollutants are left on the exterior of the snow crystals that come down as snowfall. When the snow begins to melt in spring, these pollutants are released first, so meltwater contains higher concentrations than the snow overall. Thus, the acids are released at enhanced concentrations in the early spring meltwaters [1]. This is the very time when young fish will be present in streams and they are especially sensitive to acidity.

In the 1970s, the forests of Europe were recognized to be in decline. Part of the explanation was an increasing exposure to air pollutants [2]. Additionally, acid rain had the potential to leach important mineral nutrients from soils, while in urban areas the acids could threaten building stone, although this was mostly caused by the deposition of local sulfur dioxide onto stone facades, rather than rainfall [3].

A range of strategies were used to reduce the acid rain problem, importantly, removing sulfur from flue gases of large coal burning power plants, moving to low sulfur coals or even shifting to gas as a fuel. Some ecosystems were so strongly acidified by the continuing deposit of acids that more active remedies were required, so alkali in the form of limestone were added to the waters of some strongly affected lakes.

As sulfur emitted to the atmosphere declined so did the deposited acids, but the changes had other effects. Agriculture, which had benefited from the input of sulfur began to reveal evidence of sulfur deficiency in some crops in Northern Europe [4]. Reductions in the deposit of pollutants in Europe and North America were counterbalanced by industrialization in China and India where acid rain came to be a growing concern [5].

3.3. Stratospheric Ozone Depletion

Just as the world was becoming aware of the acid rain problem, another emerged. The stratosphere, a part of the upper atmosphere (20 km near the equator, but 7 km at the poles), began to be explored with rockets and subsequently high-altitude aircrafts. Initially, this was to gather evidence about radioactive debris from nuclear tests, but it soon established the presence of sulfur compounds in the upper atmosphere.

The underlying theory of ozone (O_3) formation had been proposed by Chapman in the 1930s. The formation ozone being given as:

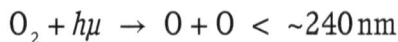

$$O_2 + h\mu \rightarrow O + O < \sim 240\,nm$$

$$O + O_2 \rightarrow O_3$$

with two key removal processes:

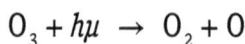

$$O_3 + h\mu \rightarrow O_2 + O$$

$$O + O_3 \rightarrow O_2 + O_2$$

But although the theory gave a good approximation to the ozone profile, it was not entirely able to explain the concentrations of ozone observed. The scientist, Paul Crutzen recognized the need for other removal pathways:

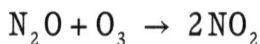

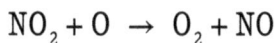

$$N_2O + O_3 \rightarrow 2NO_2$$

$$NO_2 + O \rightarrow O_2 + NO$$

$$O_3 + NO \rightarrow O_2 + NO_2$$

In the 1970s Sherwood Rowland and Mario Molina realized that increasing releases of refrigerant CFCs (represented below as $CFCl_3$) would offer additional removal pathways, where chlorine would contribute to ozone removal:

$$CFCl_3 + h\mu \rightarrow Cl + CFCl_2$$

$$O_3 + Cl \rightarrow O_2 + ClO$$

$$O + ClO \rightarrow Cl + O_2$$

and thus the idea of a human contribution to ozone depletion was born. The theory was again not complete as the expected ozone removal was not to be found. It was realized that the reaction:

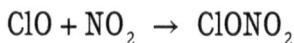

$$ClO + NO_2 \rightarrow ClONO_2$$

would effectively remove chlorine, so perhaps it might not be such a problem.

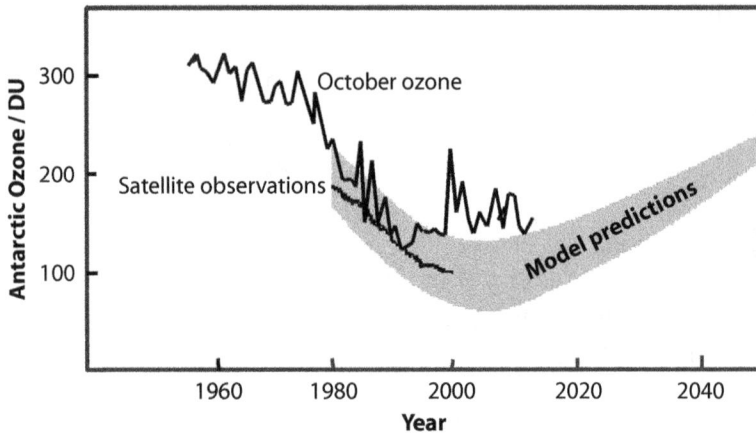

Fig. 4: Measured and Modeled Ozone in the Antarctic Stratosphere Over Time (October)

However, suddenly in the 1990s, a long series of observations of ozone over the Arctic were reanalyzed by Joe Farman of the British Antarctic Survey, Survey, showing enormous losses of ozone over the years (Fig. 4). It was subsequently established that a set of heterogeneous reactions in polar stratospheric clouds effectively enhanced the role played by chlorine in the spring stratosphere over the southern continent [6].

As a response to the threat to the ozone layer, an international treaty designed to protect it (Montreal Protocol) was formulated to encourage the replacement of substances that are responsible for ozone depletion. It was agreed on 26 August, 1987, though it has undergone many revisions. It has been remarkably successful in being able to address the threat to stratospheric ozone. However, the sensitivity of ozone to an ever-widening range of compounds has become apparent, so it has been necessary to modify the protocol. As an example, a new threat to ozone may arise through increases in CH_2Cl_2 and CH_2ClCH_2Cl in surface and upper-tropospheric air. These may be released due to a boom in the use of PVC sheet in agriculture and the construction industry [7].

3.4. *Biomass Burning, Volcanoes, and Dust*

Air pollutants may also arise from natural sources, though often mediated by humans. Forest fires can be natural occurrences, initiated by lightning, but they may also be deliberately set. Humans have used this as a method of clearing land for many millennia. Thus, forest fires represent a source of pollution that has both natural and anthropogenic components. Such fires have been common across tropical parts of South America and South East Asia, with lesser known sources in Siberia. In recent years, in the United States and Australia there have been many large-scale fires, which have affected the air pollution found in nearby cities much as they have in South East Asia for many decades [8]. These fires yield a range of gases in addition to particles, and notably after long-range transport, ozone concentrations can be enhanced further afield. However, it has been the smoke particles that have raised the greatest concern. The air pollution guidelines values in South East Asia, and more recently in Australia, have often been exceeded during times when the fires are widespread. The effect of fire smoke particles on health are less well understood than the particles more typical of cities, where particles arise from the combustion of fossil fuels used by industry and transport sources [9]. The health outcomes of forest fire smoke may be different, and there is some evidence that the effects may be long term. There is much difficulty in controlling the fires set, because for many small farmers and larger industries it is seen as an inexpensive way of clearing land and even where it is discouraged, it has been hard to stop.

A parallel problem is that of windblown dust. The most regular is the yellow dust that blows from the deserts of western China each spring [10]. Although desert dust seems a natural occurrence it can also be affected by human activities. Desertification can be the product of over utilization of water supplies, overgrazing, or crop cultivation. The most striking example may be the near loss of the Aral Sea, because so much water has been extracted. This has led to an 80% reduction in its volume, such that the high salinity and the accumulation of pollutants has caused the loss of most fish. The cumulation of organochlorine pesticides, polychlorinated biphenyls (PCBs), and dioxins has additionally compromised the health of those living in the area [11]. It is an example of a grand scheme aimed to enhance the regions agricultural output (largely as cotton), which has gone entirely wrong.

Volcanoes, despite being natural, give rise to many problems. Intense eruptions push gas and dust into the stratosphere, where it affects the radiative balance and often leads to cooler global surface temperatures. Volcanic ash in the upper troposphere can cause problems for aircraft, as the mineral particles can damage sensitive surfaces and lead to engine failure, so airspace is restricted after major eruptions. Perhaps most notable was the effect of the eruption of Eyjafjallajökull in 2011, which disrupted air traffic across the North Atlantic and led to many travelers becoming stranded [12]. Additionally, volcanic emissions can spread regionally and cause health problems in locations such as Hawaii [13], often termed *vog* (a combination of the words for volcano and fog).

3.5. *Oil Pollution*

Oil is a key fuel that has powered our enormous growth for more than half a century. Supplies are typically located far from where they are needed, the Arabian Peninsular being a key location for oil wells. In general, crude oil and its refined petroleum products are transported by ship, although liquid natural gas can be conveniently transported by pipeline. Losses from ships have been reduced over the years with more care with procedures to clean tanks. However, oil field and shipping accidents have frequently led to large releases of oil. When the Torrey Canyon ran aground off Cornwall in 1967, some 0.14×10^6 L of oil was lost. More recently the Deepwater Horizon drilling rig explosion in the Gulf of Mexico, April 20, 2010, caused a sea floor release of oil that amounted to 9.5×10^6 Ld^{-1} [14]. The amount of oil lost from 15 major oil spills is shown in Fig. 5, but of course the very large number of minor spills cannot be included. Larger spills are rare and unique events as is typical of many environmental releases, which would show a similar relationship.

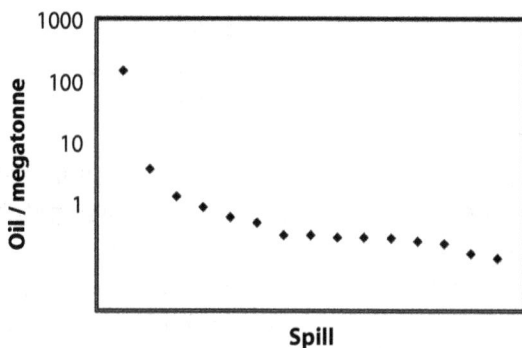

Fig. 5: Size of the 15 Largest Oil Spills (i) Kuwaiti Oil Fires, (ii) Kuwaiti Oil Lakes, (iii) Lakeview Gusher, (iv) Gulf War oil spill, (v) Deepwater Horizon, (vi) Ixtoc I, (vii) Atlantic Empress/Aegean Captain, (viii) Fergana Valley, (ix) Nowruz Field Platform, (x) ABT Summer, (xi) Castillo de Bellver, (xii) Amoco Cadiz, (xiii) Taylor Energy, (xiv) MT Haven, (xv) Torrey Canyon

Oil contains hydrocarbons such as alkanes along with aromatic hydrocarbons, and in particular the polyaromatic hydrocarbons (PAHs). However, much of the material in oil is *unresolved complex matter* [14]. Oil pollution causes increased deaths among seabirds and marine mammals, but this arises not so much through chemical toxicity, but more from the physical properties of oil. These enable it to dissolve fat layers, which isolate, e.g., a seabird's plumage or otter's fur. Mortality is the result of drowning or hypothermia [15], while when oil washes ashore it damages intertidal ecosystems.

3.6. Plastic Pollution

The problem of plastics in the ocean achieved prominence over the last few years (~2018), but plastics had been observed in the oceans for more than 50 years [16]. Currently, we produce almost 400 million ton of plastic each year and some 275 million ton appears as plastic waste. Some 5–13 million ton is deposited in the sea through storm water discharge, littering, loss from solid waste disposal, along with some from sea-going vessels, and the offshore oil and gas industry.

Plastic floats over great distances and accumulates in ocean gyres, one being called the Great Pacific garbage patch, which results in a large amount being deposited on remote islands, where it presents a hazard to marine life and seabirds. Although plastic is durable it can degrade through exposure to ultraviolet (UV) irradiation or via biological attack, perhaps more rapidly than the century-long lifetimes once claimed. Recently Royer et al. [17] have shown that some common plastics produce two greenhouse gases, methane and ethylene, on exposure to sunlight.

Smaller microplastics (pieces in the size range 2–5 mm), sometimes called *mermaids' tears*, have recently become of interest. These can result from the abrasion of friable plastics (e.g., polystyrene) or solar irradiation of plastics, such as polypropylene. Figure 6 shows a scanning electron

Fig. 6: A Polypropylene Bead Degraded in Sunlight and Water

micrograph of the fractured surface of a polypropylene bead, which has been exposed to sunlight for many months. The cracks slowly reduce the plastic to tiny flakes, which can be ingested by marine animals and zooplankton. Styrofoam fragments can be found in piles on beaches after storms. In seawater these ubiquitous polystyrenes degrade to their styrene oligomers, to be found in North-West Pacific at 0.17–4.26 µg L^{-1} [18]. As carcinogens the compounds represent a potential hazard.

The weathered surfaces of microplastics allow them to absorb toxic substances from the water [19], thus concentrating harmful substances making the particles more toxic to marine fauna. In the smallest size range, plastic particles or sometimes fibers can become airborne. These have been detected in human lung biopsies, where it is feared that they can induce disease in susceptible individuals [20].

4.0   Local Pollution

Pollution raises special concern at a local level, where its effects are readily apparent and generate immediate and noticeable concern. There are often distinctive approaches to control that can be administered locally and lead to designation in planning requirements, the inspection of everything from vehicles to factories and demand for stricter compliance when new equipment is purchased.

4.1.   Urban Air Pollution

Poor air quality in our cities is one of the most frequent concerns of the large numbers of people who now dwell there. Cities have been polluted for a long time [21]. Complaints were frequent in ancient Rome where the administrator Frontinus described the city's unwholesome (*gravioris caeli*) and suspect air (*infamis aer*) were harmful. However, the problem grew worse when fossil fuels, particularly coal, started to be used. In London this was a problem by the late 13th century when Royal proclamations tried to restrict its use. This was not so much because they knew about the harmful components of coal smoke, but more because the strange smell suggested a potential risk, as centuries ago it was believed that disease arose from bad airs (miasmas).

Combustion of fuels, which are typically hydrocarbons (C_xH_y), would produce water and carbon dioxide on complete combustion. Combustion may not be complete and particularly where the supply of air is limited, then incomplete oxidation means that carbon monoxide or soot is produced. In some cases, the fuel can pyrolize and result in the formation of larger organic dehydrogenated molecules, such as polycyclic aromatic hydrocarbons (PAHs).

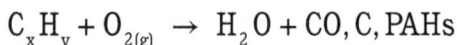

$$C_xH_y + O_{2(g)} \rightarrow H_2O + CO, C, PAHs$$

PAHs are large organic molecules usually dominated by six-membered benzene rings. The smaller PAHs, such as the parent compound benzene or naphthalene, are found in the gas phase, but many are semivolatiles, so they are present as particles or bound to particles in the atmosphere. They are especially worrying as many are carcinogenic.

Fuels contains sulfur as an impurity, and as noticed in the discussion of acid rain, this leads to sulfur dioxide. Concentrations tend to be high in coal-burning cities, but some heavy fuel oils can also contain substantial amounts of sulfur. Chlorides and metals can be found in the ash, both from oil and coal. In the past, lead has been added to petrol, but this has ceased and has led to a reduction in the potential exposure to lead in our cities. Good news as it is such a toxic element.

The discussion thus far is about pollutants that originate directly from release, here from an exhaust pipe or a factory chimney. These are termed primary pollutants and are easily recognized as the plumes from tall stacks. However, by the mid-20th century it became evident that this simple view of air pollutants did not describe the pollution that was especially bad in Los Angeles. The lettuce crop was damaged by some unknown pollutant, so after studies by the biochemist Haagen-Smit, it was recognized that this was caused by ozone. This was produced from reactions related to volatile organic compounds (automotive vapors) in the sunlit atmosphere. Thus the concept of secondary pollutants was established. Through the 1950s the mechanism was unraveled by Leighton and other workers [22].

The ozone in photochemical smog comes about through the photostationary state, which can be written as a kind of equilibrium:

$$O_3 + NO \leftrightarrow NO_2 + O_2 \, (+\text{light})$$

The forward reaction produces the brown colored gas nitrogen dioxide (NO_2). In sunlight this can be reversed to give back ozone and nitric oxide. If the air has other pollutants such as hydrocarbons, the volatile organic compounds that would evaporate from petrol or combustion process, then another route to ozone formation becomes available. Representing the reactions in terms of the simplest hydrocarbon methane, it is possible to write a sequence of reactions that follow the attack by an OH radical:

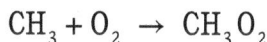

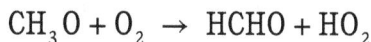

$$CH_4 + OH \rightarrow CH_3 + H_2O$$

$$CH_3 + O_2 \rightarrow CH_3O_2$$

$$CH_3O_2 + NO \rightarrow CH_3O + NO_2$$

$$CH_3O + O_2 \rightarrow HCHO + HO_2$$

$$HO_2 + NO \rightarrow OH + NO_2$$

The OH radical remerges at the end so it a cycle that has essentially allowed oxidation of the hydrocarbon to an aldehyde (or a ketone) and gives nitrogen dioxide, as summed below:

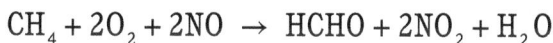

$$CH_4 + 2O_2 + 2NO \rightarrow HCHO + 2NO_2 + H_2O$$

This NO_2 can now shift the photostationary state to the left producing the ozone, which characterizes photochemical smog. Along with this ozone and oxidized organic compounds,

276 Chapter 7: Sustaining Our Air and Water

potentially irritants and carcinogens, there are fine particles or secondary organic aerosols. All further threats to health.

4.2. Urban and Rural Runoff

Land areas are a source for water pollution. In cities this appears during heavy rainfall as runoff from hard surfaces, such as buildings and roads. These surfaces have little capacity to absorb water, so the flow exhibits a very sharp hydrograph; rain that falls immediately flows over the surface [23]. This flow can be very rapid, so not only does it carry pollutants in solution or as a suspension, it can be aggressive enough to carry larger solid fragments. Indeed, at times the fragments can be very large, with intense storms moving objects such as cars along with the flood waters. Though it is the smaller material that has a great potential to cause environmental problems. Typically, runoff from storms will carry rubber fragments, heavy metals, salts, detergents, and sewage from over-flowing systems. Urban stormwater has become an increasingly important source of nitrogen and phosphorus to receiving waters, raising the potential for eutrophication [24], especially where storm water has overwhelmed the sewage treatment system. There are differences depending on the sources of runoff: street runoff shows high levels of suspended solids, oxygen consuming chemicals (COD), and hydrocarbon loads, while roof runoff has high concentrations of heavy metals [25].

Rural areas will have broader hydrographs as the water flows over more absorbent surfaces before it discharges into watercourses. During floods, such water will carry sediments washed from the land, which means flood waters take on a color typically dependent on the local soils. Such fine mud generally causes only minor environmental problems and in some cases this material, as was the case in the annual Nile floods, contributes to the agricultural fertility of the area where the sediment deposits. Pollutants incorporated into the rural runoff may include pesticides, veterinary drugs (excreted by farm animals), urine, feces, and fertilizers.

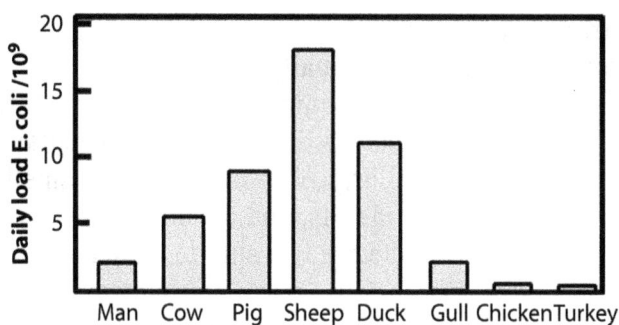

Fig. 7: Daily Release of *E. Coli* From Humans and Animals

Agricultural communities have traditionally developed close to rivers, streams, or other readily accessible sources of freshwater. However, in the late 20th century, intensive animal husbandry in the developed world led to increasingly concentrated loads of animal feces from agriculture and consequently incidents of significant organic pollution to river water from agricultural practices. The presence of *Escherichia coli* concentrations found in the feces of various animals are given in Fig. 7. Although, inputs from diffuse and point sources of pollution were not quantified, point sources (such as channeled runoff from farmyards) seem also likely to be significant. [26] reviewed the fecal pollution from storm water runoff from animal feedlots in the United States, where thousands of cattle were confined and found that it equated to the discharge of raw wastewater from towns that approached populations of 10,000 [27].

4.3. Factories, Sewage Works, and Large-Point Sources

Large-scale industrial developments discharge waste to both the atmosphere and rivers. Partly this arises because they require a large energy input, so combustion generated primary pollutants can be released in large quantities, typically NO_2 and SO_2. The cement industry is notable as it accounts for 8% of global carbon dioxide emissions, part of this is released from flue combustion required to heat the limestone, but part also comes from the chemical transformation of the limestone, producing around 4-billion-ton carbon dioxide annually.

Thus, besides the energy requirements that characterize industry there may be emissions derive from a particular set of industrial processes. Metallurgical production from smelters can release large quantities of metals, and in some cases such as nickel production, where the ore is a sulfide, much sulfur dioxide.

$$Ni_2S_3 + 4O_2 \rightarrow 2NiO + 3SO_2$$

Brickmaking is often characterized by emissions of hydrogen fluoride and thiols (organosulfides taking the form R–SH), which smell like burning rubber. Paper and pulp manufacture frequently uses large quantities of chlorine and leads to the release of hexachlorobenzene, chlorinated dioxins, and furans along with a range of sulfides: hydrogen sulfide, methyl sulfide, and dimethyl sulfide in kraft mills.

The river pollution can be as point discharges from factories, so here waste can arise from the production of dyes and sulfides and phenols, with chromium from the textile manufacture. One ton of paper requires 75,000 liters of water, so the industry is often a major polluter. Wastewater from paper production can contain suspended solids, nutrients, and dissolved organic matter, such as lignin, because of the nature of the processes it frequently contains organochlorine and organosulfur compounds. In food production, vegetable washing generates waters with high loads of particulate matter, surfactants (detergents), and some dissolved organic matter. Animal slaughter additionally releases organic nitrogen and ammonia.

Sewage works can be a special source of pollution to rivers. Many problems along with the pollutants released as effluent from sewage works can be a special problem. Early legislation dealt with the problem and established standards and tests to be applied to examine sewage and sewage effluents, introducing the concept of BOD (biochemical oxygen demand). However, the first attempt at establishing effluent standards came in 1908. Preliminary, empirical standards were proposed, based on an analysis of data from existing plants and established what would be expected of a well-operated treatment works. Additionally, sewage treatment had to cope with potential releases of toxic metals, biological pollutants, organic compounds in addition to consumer products such as sunblocks and pharmaceuticals, which are not always destroyed during treatment.

Liquid waste also seeps out from waste facilities, in particular landfills, as a leachate. Typically landfill sites will not have large amounts of clinical or chemical waste, but more often waste from municipal, commercial, and industrial sources. Rainwater percolates down through the waste material and any oxygen tends to get used during the oxidation of organic materials reducing the volume of the waste and generating a range of simple organic compounds (alcohols, acids, aldehydes, short-chain sugars, etc.), which can dissolve. The leachate will typically include sulfate, chloride anions along with the cations of iron, aluminum, zinc, and ammonia, and some toxic metals such as cadmium, chromium, copper, lead, zinc, and nickel. As the leachate evolves it becomes anoxic. Anaerobic digestion by microbes can lead to the production of landfill gas, which is largely methane and carbon dioxide, although if there are large volumes of building waste with gypsum plaster (calcium sulfate) considerable quantities of hydrogen sulfide can be generated.

Power stations, paper mills, and other industries are often found along riverbanks to allow ready access to water used in their operation. The water is also used as a coolant, but this results in the discharge of warm water into rivers, streams, and lakes. Although rapid changes in temperature can kill fish through a thermal shock, increases of 2–3°C can affect the breeding of some aquatic insects and even small temperature changes of just 1–2°C can alter the ecology.

The Pearl River Delta that sits within the Guangdong-Hong Kong-Macau Greater Bay Area is among the fastest growing regions of China. It has recently seen as one of East Asia's pollution hot spots, with major impacts spilling into the South China Sea. Researchers have examined the parameters for water quality variations between the river catchments [28]. In the north, pollution was related to organic constituents (DO, dissolved oxygen, and COD, chemical oxygen demand), inorganic nutrients

(NH$_3$-N and TP, total particulate concentration), and mercury, while in the east, biological oxygen demand and inorganic nutrients (NH$_3$-N and TP) In the west, mainly related to organic-related parameters (COD) and inorganic nutrients (NH$_3$-N and TP), illustrating considerable range across the region.

Estuaries and river deltas are highly productive ecosystems, so humans often choose to live along their margins. This means that estuaries are susceptible to pollutants both from their edges and through pollution brought downstream by the rivers, which run into the estuary. This makes them a receptor for herbicides, pesticides from the agricultural hinterland and sewage, heavy metals and detergents. They can readily become threatened ecosystems.

5.0 Mitigation

5.1. *Legislative and Technical Approaches*

Pollution can be reduced through legal and technical approaches, although these could understandably lead to social change. Legislative mechanisms, which in some cases are international rather than national, have long been favored, but require the availability of technically feasible solutions. The success of the *Montreal Protocol* came about because it was relatively easy to find substitutes for the ozone depleting CFCs used in air conditioning, refrigeration, and as aerosol spray propellants. Although replacements went through several different forms, it was relatively easy to find those which would be less harmful overall. Nevertheless, it is always important to consider when replacing one environmentally threatening material with another, it may simply make the threat subtler, so any substitute is to be adopted with caution.

Environmental regulations are widely seen as an effective way to ensure cleaner air and water. These regulations may be based on controlling emissions to the atmosphere or water bodies. The National Environment Agency of the Singapore Government sets limits for trade effluent discharge to watercourses; in the case of copper this is 0.1 mg/L [29]. Emissions are also set, e.g., for air pollutants from vehicles, so back in 1978 when Japan set vehicle limits to the *hot-start* emissions of carbon monoxide, these were to be less than 0.25 g per kilometer. The standard is neatly tuned to the operation of the emission source, yet is also useful because it gives clear guidance on what is to be achieved by a manufacturer or an industry, or the desired target during maintenance of a vehicle. However, this may not always be appropriate as emissions do not necessarily represent human or ecosystem exposure, as limiting this is usually the goal of regulation. This is reflected in an increasing interest in exposure science. Thoughtful approaches are needed because it is evident that there is a worry that pollutant emissions to the atmosphere are not linearly related to their concentrations because the ones in the environment undergo chemical transformation or removal.

In the case of many pollutants, it is necessary to set allowable values for the concentrations at which they are found in the environment. In the atmosphere, it is typical to set limits or guideline values to the concentrations of key pollutants. The *Air Quality Standards* [30] within the European Union set the limit value for nitrogen dioxide (NO_2) at 200 µg/m^3. However, we should note this is to apply over a defined period because the observed health impacts associated with various pollutants occur over different exposure times. The specified concentrations for a pollutant can be rather subtle. In the case of NO_2 the limit value applies for an averaging period of 1 hour with 18 exceedances permitted each year. Along with this there is often a *time of compliance*, so with NO_2 there was required that the standard was to be met on January 1, 2010.

Naturally, an improved environment cannot come about simply because laws have been passed; compliance and enforcement can be difficult. It may seem that pollution control is just about bringing a set of regulations onto the statute books, but it is the application of these that ultimately leads to improved environmental quality, and this is far from automatic. Enforcement requires good analytical capability to establish the extent to which guidelines are met. It is a common concern that enforcement is slow and lacks any teeth, with the financial penalties so small that they fail to deter the polluter. Despite this, some troublesome pollutants have been substantially reduced over time. Nevertheless, the nature of pollution has changed. Considering air pollution there is a shift from large stationary sources burning coal and emitting smoke and sulfur dioxide to numerous mobile sources (i.e., vehicles) using liquid fuels and releasing nitrogen oxides, volatile organic compounds, and fine particles. It is not always that pollution gets better or worse, at times it merely seems that its character changes.

Legislative and technical approaches are often seen as part of a command and control strategy, and although potentially effective are not the only route to a better environment. Some changes are more difficult to implement, very costly and have broad impacts on our lives. The emissions of gases such as carbon dioxide seem to have an undeniable link to greenhouse warming, yet it has been a problem that has failed to achieve consensus either about its existence or the appropriate solution. This is partly because it can only be solved in ways that are likely to change our lives and livelihoods. The key solution is to reduce emissions of carbon dioxide, which involves a major shift away from traditional approaches to generating energy, i.e., burning fossil fuels. While great strides have been made in developing renewable or perhaps small-scale nuclear generation of energy, they are not always easy to adopt, so remain a challenge. Solar-powered aircraft are in their infancy and although sailing ships would seem a century-old solution to maritime emissions, their speeds are variable in an age that demands on-time delivery.

5.2. *Engineering Solutions*

Engineering design can play a key role in lowering the emission of pollutants. This can begin with the planning that chooses suitable site locations and design. Design of factories and related equipment is especially relevant to major industries such as power plants, factories, or infrastructure such as harbor facilities and airports. Emission control from large plant often involves the need for dust and fly ash collection. This requires systems such as cyclone separators, which can be useful for larger particles such as sawdust, baghouses with their fabric filter, and electrostatic precipitators. Scrubbers can also remove particles: (a) wet-scrubbers, which spray water into a column to remove water-soluble gases and (b) dry-scrubbing systems have the advantage that the exhaust gas stream is not saturated with moisture. Exhausts also need to be treated on a smaller scale such as in automobiles. Here removal can make use of catalytic converters, but they demand to be at a given temperature and have the right level of oxygen in the gas stream.

Water pollution is typically controlled through treatment of the wastewater, prior to release. Industrial wastewater treatment can be associated with a particular facility. In cities there are sewage works to treat municipal wastewater—facilities that are often large and involve tanks for settling and aeration and methods to handle the sludge. Sometimes systems have to be more elaborate if the water discharged needs to be of the highest quality. In such situations tertiary treatment is needed, which may involve an exacting level of chemical or particle removal.

Maintenance and restoration of infrastructure such as that in sewerage systems is increasingly important as in many cities these systems have become more than a century old [31]. Large-scale planning for airports or transport systems need to take a wide-ranging approach, and consider that flood control, parking, or link roads will affect the local environment. Planning and risk assessment are essential at an early stage.

Solid waste is often sent to landfills, but over time this can release gases such as methane. In some cases, the landfill gas can provide some local source of energy, though the quantities are relatively small. Geomembranes or engineered clay linings to landfills aim to prevent leakage, especially of water that has drained through the site as a leachate. The loss of this, which can be highly polluting, can be a considerable worry at landfills.

5.3. Soft-Engineered Solutions

Often engineering is envisaged in terms of large structures and built facilities, but it can also provide softer solutions, which are more related to planning and design. Increasingly street runoff is controlled by soakaways filled with rubble or green verges of vegetation along roads. These encourage rainwater infiltration, reducing storm water flooding, but also reduce pollution through bioretention. Such infiltration-based methods are widely used because of their ability to improve water quality and the hydrologic condition of the urban landscape. Bioretention systems are effective at removing a range of pollutants, including suspended solids, heavy metals, phosphorus, oil and grease, along with fecal coliform bacteria. Nitrogen removal performance, though has been highly variable, with reported results ranging from as high as 60% removal to net nitrogen export [24].

Urban layout can be carefully designed through zoning, which places compatible uses in proximity. Noisy and polluting roads can be diverted away from school and playgrounds. Bicycle trails and footpaths can encourage cyclists and pedestrians to adopt less heavily trafficked routes, reducing air pollution exposure. There can be social features that affect vehicle use, such as encouraging students to walk to school (the walking bus) as an approach to air pollution control. Urban parks can offer buffer zones, and also use vegetation and other have the park infrastructure to encourage visitors to frequent less polluted areas [32].

5.3. *Economic Instruments*

Control of pollution is often seen as technology-based and performance-based, yet there are fiscal approaches. Economists perceive pollution as a "market failure," which arises because "polluters" see little cost to the release of pollutants, so there is no incentive to curb emissions. This can be changed by imposing charges, fees, or taxes. Thus, a price is paid when pollution results from the production, use, or disposal of goods. In some cases, this may take the form of deposit–refund schemes. Another approach is to have Tradable Permits for pollutant emissions. If these are traded it is possible to add a levy that makes the pollution more expensive after each transaction, thus pressuring reduction over time.

Often pollution can be thought of in terms of *Cost–Benefit Analysis* showing that while expenditures may be large, the benefits are often even larger. However, a continuing problem is that such analyses may not account for the issue that the costs and benefits are often distributed quite differently. This may be accounted for in other ways and often it is necessary to consider concepts such as the willingness-to-pay and assess weightings for different individuals.

Another popular consideration in the economics of clean air and water is the Environmental Kuznets Curve. This curve can be encapsulated within the idea than in the early stages of societal or industrial growth pollution increases, but it reaches a point where societies are no longer willing to accept a degraded environment as part of further growth. There is no fundamental reason why pollution should follow such a profile, yet it is widely observed. It seems more evident where local pollutants can be reduced such that any resultant improvements are locally evident. It is more difficult to imagine that this would happen with global pollutants, such as carbon dioxide emissions when a single community or nation is not likely to make a great difference as immediately observable outcomes.

5.4. *Sociological Change*

Environmental issues have increasingly engaged the public. The acid rain issue dominated environmental thinking of the 1980s. It was one of the first truly global environmental problems that drew a worldwide audience. Although, in some ways this attention came at a point when many of the issues relating to acid rain were well understood and emissions already in decline. Acid rain catalyzed the environmental views of a generation and became embedded in school geography texts for many decades. The desire for international control meant it was debated at the 1972 United Nations meeting in Stockholm and led to the *Convention on Long-Range Transboundary Air Pollution* of 1979. Much effort was required to reach agreement in 1993, to reduce sulfur emissions or transboundary fluxes by 30%. Such international protocols have also been followed to reduce ozone depleting substances (with considerable success) and greenhouse gases (with far less success).

Social attitudes not only toward plastic, but also clothing, vehicles, and choice of transport have gradually changed. The enormous consumption of plastics and the waste it creates, clearly presents us with environmental and health risks; so what is the solution here? As always reducing the production and use seems important. Abandoning single-use items made of plastic, i.e., moving away from drinking straws or avoiding plastic shopping bags can help. It is consumer behavior that can drive this, but fiscal incentives are also effective, witness the charge for shopping bags in many countries. Recycling needs to increase. In the United States only 6% of the plastic in municipal solid waste is recycled. There is enormous room for change; considering the world uses a million plastic drink bottles a minute, yet few are recycled. Some plastic can be reused, some can be respun into fabrics, deposit schemes, and charges for single-use are often effective. The pressure may in the end be driven by the consumer and our inventiveness, so we can all contribute to creating an environment freed from the risk of waste plastic.

Engaging the public is important. There are examples where the public have led change as in the reduction of shipping emissions or in concerns over the ineffectiveness of short-term solutions to pollution such as closing factories during the Olympic Games or major international meetings [33] and even where the public don't lead, having them on-board is very important, as seen with regulating firework sales to reduce pollution [34].

6.0 Conclusions

Numerous environmental problems occur over a wide range of scales, which can reflect national or urban approaches to the environment. Creating regulations is important, but it is only as effective as its implementation. Technical approaches need to consider everything from the initial design through to maintenance. It is important to ensure that a policy designed to reduce pollution, doesn't make things worse or simply add a further more subtle problem that will appear later. It is also relevant to consider economic controls to pollution along with issues of social change, sometimes these softer approaches, which can engage the public will have the potential to bring about more lasting improvements.

7.0 Extra Reading

https://www.sciencedirect.com/topics/earth-and-planetary-sciences/river-pollution

References

(1) Brimblecombe P, Tranter M, Abrahams PW, et al. Relocation and preferential elution of acidic solute through the snowpack of a small, remote, high-altitude Scottish catchment. *Ann Glaciol.* 1985;7:141–147.

(2) Mueller-Dombois D. Forest decline and dieback—a global ecological problem. *Trends Ecol Evol.* 1988;3(11):310–312.

(3) Brimblecombe P, Grossi CM. Millennium-long damage to building materials in London. *Sci Total Environ.* 2009;407(4):1354–1361.

(4) Schnug E, Evans EJ. Monitoring of the sulfur supply of agricultural crops in northern Europe. *Phyton.* 1992;32(03):119–122.

(5) Brimblecombe P. Acid rain.In: Ritzer G, ed. *The Wiley-Blackwell Encyclopedia of Globalization.* Chichester: Wiley-Blackwell; 2012.

(6) Peter T, Grooß JU. Polar stratospheric clouds and sulfate aerosol particles: microphysics, denitrification and heterogeneous chemistry. In: Muller R, ed. *Stratospheric Ozone Depletion and Climate Change.* London: Royal Society of Chemistry; 2012:108–144.

(7) Oram DE, Ashfold MJ, Laube JC, et al. A growing threat to the ozone layer from short-lived anthropogenic chlorocarbons. *Atmos Chem Phys.* 2017;17(19):11929–11941.

(8) Latif MT, Othman M, Idris N, et al. Impact of regional haze towards air quality in Malaysia: a review. *Atmos Environ.* 2018;177:28–44.

(9) Liu JC, Pereira G, Uhl SA, Bravo MA, Bell ML. A systematic review of the physical health impacts from non-occupational exposure to wildfire smoke. *Environ Res.* 2015;136:120–132.

(10) Altindag DT, Baek D, Mocan N. Chinese yellow dust and Korean infant health. *Soc Sci Med*. 2017;186:78–86.

(11) Whish-Wilson P. The Aral Sea environmental health crisis. *J Rural Remote Environ Health*. 2002;1(2):29–34.

(12) B.Langman North Atlantic Volcanic Ash (2010): contemporary social vulnerability to a natural event. *Air Pollut Epis*. 2017;6:287.

(13) Businger S, Huff R, Pattantyus A et al Observing and forecasting Vog dispersion from Kīlauea volcano, Hawaii. *Bull Am Meteorol Soc*. 2015;96(10):1667–1686.

(14) Macías-Zamora JV, Ocean pollution. In: Letche T, Vallero D, eds. *Waste: A Handbook forManagement*. Amsterdam, the Netherlands: Elsevier: 2011; 265–279.

(15) Beiras R. *Marine Pollution: Sources, Fate and Effects of Pollutants in Coastal Ecosystems*. Amsterdam, the Netherlands: Elsevier; 2018.

(16) Venrick EL, Backman TW, Bartram WC, et al. Man-made objects on the surface of the central North Pacific Ocean. *Nat*. 1973;241(5387):271.

(17) Royer SJ, Ferron S, Wilson ST, Karl DM. Production of methane and ethylene from plastic in the environment. *PLoS ONE*. 2018;13(8):e0200574.

(18) Kwon BG, Amamiya K, Sato H, et al. Monitoring of styrene oligomers as indicators of polystyrene plastic pollution in the North-West Pacific Ocean. *Chemosphere*. 2017;180:500–505.

(19) Barnes DK, Galgani F, Thompson RC, Barlaz M. Accumulation and fragmentation of plastic debris in global environments. *Philos Trans R Soc B: Biol Sci*. 2009;364(1526):1985–1998.

(20) Prata JC. Airborne microplastics: consequences to human health? *Environ Pollut*. 2018;234:115–126.

(21) Brimblecombe P. *The Big Smoke (Routledge Revivals): A History of Air Pollution in London since Medieval Times*. Routledge; 2012.

(22) Brimblecombe P. Deciphering the chemistry of Los Angeles smog, 1945–1995. In: Fleming JR, Johnson A, eds. *Toxic Airs: Body, Place, Planet in Historical Perspective*. Pittsburgh: University of Pittsburgh; 2014:95–108.

(23) Walsh CJ, Fletcher TD, Burns MJ. Urban stormwater runoff: a new class of environmental flow problem. *PLOS ONE*. 2012;7(9):1–10.

(24) Li L, Davis AP. Urban stormwater runoff nitrogen composition and fate in bioretention systems. *Environ Sci Technol.* 2014;48(6):3403–3410.

(25) Gromaire-Mertz MC, Garnaud S, Gonzalez A, Chebbo G. Characterisation of urban runoff pollution in Paris. *Water Sci Technol.* 1999;39(2):1–8.

(26) Geldreich, EE, Microbiology of water. *J Water Pollut Control Fed.* 1972;44(6):1159–1172.

(27) Taylor H. Drinking water microbiology. In: Mara D, Horan N, eds. *The Handbook of Water and Wastewater Microbiology.* London, England: Academic Press; 2003:611–777.

(28) Fan X, Cui B, Zhao H, Zhang Z, Zhang H. Assessment of river water quality in Pearl River Delta using multivariate statistical techniques. *Procedia Environ Sci.* 2010;2:1220–1234.

(29) NEA. Limits for trade effluent discharge to watercourse or controlled watercourse. 2020. https://www.nea.gov.sg/our-services/pollution-control/water-quality/allowable-limits-for-trade-effluent-discharge-to-watercourse-or-controlled-watercourse.

(30) European Commission. Air Quality Standards. 2019.https://ec.europa.eu/environment/air/quality/standards.htm.

(31) Read GF, Vickridge I, eds. *Sewers: Repair and Renovation.* Oxford: Butterworth-Heinemann;2004:1–19.

(32) Xing Y, Brimblecombe P. Traffic-derived noise, air pollution and urban park design. *J Urban Des.* 2020:25(5):609–625.

(33) Brimblecombe P, Zong H. Citizen perception of APEC blue and air pollution management. *Atmos Environ.* 2019;214:116853.

(34) Lai Y, Brimblecombe P. Regulatory effects on particulate pollution in the early hours of Chinese New Year, 2015. Environ Monit Assess. 2017;189(9):467.

CHAPTER 8
"Living" Coastal Protection and Nearshore Biodiversity Reclamation

Loke Ming Chou

Tropical Marine Science Institute, National University of Singapore, Singapore

Abstract

Coastal urbanization has resulted in profound and permanent environmental change. The Seaward shift of shorelines protected by solid seawalls eliminated the entire biologically productive intertidal zone leading to loss of valuable ecosystem services and leaving almost no opportunity to restore habitats. Artificial structures of the urban waterfront can support biodiversity but have severe space constraints. Ecological engineering of the seaward face of seawalls, which literally represents a highly compressed intertidal zone, can increase its biodiversity supporting capacity to a limited extent but cannot accommodate habitat regeneration. Coastal defence against sea level rise is imperative and seawalls remain as the immediate barrier. Instead of simply considering how to make a "dead" seawall higher and stronger, it will be prudent to think of transforming it into a "living" seawall or better, an eco-engineered zone that incorporates a complex of intertidal pools and lagoons, in which coastal and nearshore habitats can be restored. This nearshore biodiversity reclamation approach will help to restore ecosystem services over the long term and improve coastal area sustainability.

Keywords: Coastal urbanization, nearshore biodiversity, ecological engineering

1.0   Introduction

An estimated 40% of the world's population is settled in the coastal area, with many living in large coastal cities. Urban coastlines are typically modified by hard engineering to straighten them, create more coastal land that directly abuts the sea, and allow more direct access between land and sea. The natural undulating coastline is now replaced by a seawall that shifts the coastline seawards, obliterating the intertidal zone with a permanent wall that abruptly separates land from the sea. Like a protective armor, the seawall protects the land that it holds from being eroded by high-energy waves. This defence enables land commonly filled over the original intertidal zone to consolidate and be sufficiently compacted to support high-rise buildings and built structures typical of urban coastal cities (Fig. 1).

Fig. 1: Buildings and Ports Dominate the Urban Coastline

Defending the coastline against strong waves that erode shores has been conventionally accomplished by constructing a solid wall or revetment. Further reduction of wave energy hitting the revetment is achieved with breakwaters installed further offshore or groynes extended seaward from the shore. In nature, such protection is provided by coastal and nearshore habitats such as kelp forests and seagrasses in temperate areas, and mangroves, coral reefs, and again seagrasses in the tropics, a service that they provide at no cost to society. A simple valuation of such a service is how much it would cost to construct a revetment, a necessary action if degraded habitats can no longer provide. What is commonly not factored in the valuation are all the other important ecosystem services such as carbon sequestration, food production, and nursery function that the habitats provide in perpetuity if adequately managed. The value amounts to many times more than the cost of the seawall construction.

Singapore is a highly urbanized coastal city with its land area increased 24% by land reclamation since the 1960s. More than 60% of its coastline is lined by seawalls [1] and the drastic reduction in the extent of original coastal habitats has been documented [2]. The threat of rising sea levels necessitates the elevation of existing seawalls and their further strengthening against increasingly frequent extreme weather events. Coastal defence is of paramount importance to the survivability of low-lying coastal cities and Singapore is committed to protecting itself from the rising sea [3].

The opportunity is presented to consider a response that incorporates and enhances habitats and biodiversity so that seawalls that once functioned as a single solid non-living line of defence has now become a zone of reclaimed habitats that support rich coastal biodiversity. In designing such a zone, the main emphasis is reclamation of coastal and nearshore habitats beyond primary seawalls. Existing seawalls can be modified to increase surface area and offer conditions favoring biodiversity. Artificial structures in the urban waterfront also offer opportunities to increase biodiversity, but these interventions remain limited by the space available for colonization of epibiota and the conditions that promote the development of habitats rather than just communities.

At the same time, biodiversity can be enhanced through interventions that include active restoration, biodiversity conservation, and effective management measures against pollution and water quality deterioration. Such a zone requires a mix of civil and ecological engineering that restores habitats and improves the level of ecosystem services. This approach will help to sustain coastal and marine biodiversity in the longer term and address the United Nation's Sustainable Development Goal 14 (Life below water).

2.0   Nearshore Biodiversity

Sea-shores mark the transition between land and sea but a clear-cut boundary cannot be established in this area of changing tides. The intertidal zone (Fig. 2) is literally land when exposed at low-tide and sea when submerged by high tide. This temporally dynamic change usually twice a day presents an unclear situation to management agencies tasked with addressing impacts that are specifically land-based or sea-based. The situation presents a real administrative challenge unless an integrated coastal management framework has been adopted, which does provide the necessary coordination of responses by relevant agencies [4]. Coastal reclamation to increase the availability of land and accessibility to the sea means a drastic physical transformation of natural shores to a human-engineered waterfront that does away with natural shore habitats (Fig. 3).

Fig. 2: Vast Expanse of Intertidal Area With Rich Biodiversity Exposed at Low Tide

Fig. 3: Rockwall Protection of an Urban Waterfront

Shores slope to sea down a steep or gentle gradient. The profile influences the extent of the intertidal zone. It is more extensive on gentle-sloping shores than steep ones. Shifting tides make the intertidal zone a very challenging one for inhabiting species. Flora and fauna need to adapt to the frequent contrasting environmental conditions as the zone is subjected to the cyclical high-tide submergence and low-tide exposure. Intertidal species must find ways to avoid being dislodged by crashing waves of the incoming tide and abrasion by tiny sand particles zipping about in all directions. Prevention against desiccation or excessive loss of body moisture is important when exposed to air at low tide where wind and daytime solar radiation compound the problem. At the other extreme, heavy rainfall can induce acute salinity changes that will have a serious impact especially on species that lack osmoregulatory capability.

Intertidal species, particularly the sessile and sedentary ones, have to cope with marine predators that come in with the tide and terrestrial predators as the tide ebbs. Tidal oscillation makes the intertidal zone an exceedingly harsh environment, but despite the challenges, intertidal biodiversity flourishes because this land–sea interface is constantly enriched with nutrients from both, land and sea. At the same time, the shallow coastal waters receive abundant sunlight during the day to sustain primary production. Adequate flushing and the constant supply of nutrients, oxygen, and solar energy make the intertidal zone a highly productive area that supports rich biodiversity.

Local coastal features together with runoff and sediment discharge from land, and nearshore hydrodynamic circulation patterns influence the development of specialized coastal habitats, such as tropical mangroves, seagrasses, and coral reefs. These habitats are recognized for their rich biodiversity and high levels of ecosystem services. Other seashore habitats such as sandy, rocky, and muddy shores, algal beds, together with estuaries add to the diversity of coastal types and variety of ecosystem services. Biodiversity thrives not only on the substrate surface but also beneath, a suitable niche for burrowing species. Apart from supporting biodiversity, these habitats function as nursery areas for larvae and juveniles of many offshore species including commercially important ones.

Food and materials obtained from coastal habitats have supported coastal communities since early humankind. Habitats especially seagrasses and mangroves sequester carbon efficiently and help to bind sediment. Together with the coral reef habitat, they effectively prevent shore erosion by buffering the full force of incoming waves. At the same time, natural coastal habitats are valuable assets that keep attracting visitors and can support sustainable coastal tourism. Management is necessary to keep the habitats healthy so that society can benefit from the high levels of ecosystem services that are available without any charge.

The protective function of coasts by habitats such as temperate kelp forests [5,6] and tropical mangroves [7,8], coral reefs [9,10], and seagrasses [11,12], is widely documented as one of their many important ecosystem services. Recognition of this service heightened in the aftermath of the 2004 Asian tsunami [13–15] where huge waves surged less inland in areas behind intact habitats. The healthy habitats acted as natural barriers by reducing the full force of the waves and remained intact. Degraded habitats were ineffective at absorbing any of the wave energy allowing them to wreck devastation far inland.

Maintaining original habitats and not losing all to urbanization is an investment against future threats as it is crucial to conserve the genetic diversity needed for restoration initiatives. Despite the pace of coastal development in Singapore, many efforts have been in place to conserve coastal and nearshore habitats. One example is the preservation of coastal habitats in the development of a marine landfill, which became operational in 1999. A 7 km perimeter rock bund surrounding 3.5 square km of the sea was constructed and connected to the offshore island of Pulau Semakau. The enclosed space has the capacity to receive incinerated waste until 2035.

Mangroves on the island's eastern shore where the rock bund connected to were cleared. Compensating for this loss, two large adjacent areas totaling 13.6 ha were demarcated for the regeneration of mangroves using seedlings [16]. Sediment screens were deployed to protect coral reefs and seagrasses on the western shore from sediment generated by the construction of the perimeter bund [17]. These measures helped to retain the island's mangrove, coral reef, and seagrass habitats including the two plots of regenerated mangroves, which have matured. All have remained healthy and biodiversity-rich. The offshore landfill today functions additionally as a recreational destination for nature lovers [18]. The project demonstrated the compatibility between coastal development and habitat preservation.

The second example deals with a 92 ha mangrove habitat on the northern offshore island of Pulau Tekong. Increased shipping traffic through a narrow channel exposed the mangroves along 1.65 km of the northern shore to stronger wave energy that eroded sediment from under the mangrove roots. This resulted in tree toppling and loss of sediment holding ability at an estimated rate of 1 ha/year.

To protect and rehabilitate the mangrove, a combination of hard and soft engineering was adopted [19]. Biodegradable sacks filled with mud were packed in the cavities formed by eroded sections of the shoreline. Saplings were planted in the intertidal area where rocks of varying sizes were placed to provide better protection against waves and support anchorage. Fast-growing mangrove species were planted closer to shore while species more tolerant of strong waves and high tides were positioned further out. Two rows of bakau (hardy mangrove wood) poles were embedded in the seabed parallel to shore beyond the replanted saplings to dissipate wave energy and generate conditions more favorable to sapling growth and sediment deposition. Under urbanization pressure, it would have been more convenient to do away with the mangrove and protect the shore from erosion with a concrete seawall, but that would have meant permanent loss of a habitat together with its provision of ecosystem services.

3.0 Urbanization Impacts on Nearshore Biodiversity

Effective conservation of coastal habitats maximizes ecosystem services that benefit and sustain coastal communities. However, economic expansion and population growth have driven rapid coastal urbanization resulting in coastal use planning that is often detrimental to coastal habitats. Degradation of the coastal environment is a clear indication of coastal management that does not prioritize biodiversity benefits [20]. Enhancing biodiversity of the urban waterfront is an onerous challenge as it will involve an all-of-government planning approach with focused priority on marine biodiversity and ecology [21]. The difficulty is compounded without an integrated coastal management governance system in place [4].

Great pressure is exerted by the high proportion of population living within 100 km from the coast. An estimated 40% of the world's population is concentrated in the coastal area but in East Asia, 77% of the population live along the coast [22]. Those in rural communities remain dependent on dwindling resources from coastal habitats degraded by over-exploitation and declining environmental quality, while those in urban townships simply see disappearance of habitats from massive coastal use change.

Degradation and loss of coastal habitats left shores exposed to coastal erosion, necessitating the construction of protective seawalls. Natural habitats were also cleared or buried under coastal land reclamation sites commonly armored by seawalls. These walls are effective for protecting newly-formed seafronts from being washed away but the drastic conversion of the intertidal zone from a wide gentle-sloping shore to a narrow band along a steep solid wall effectively eradicated the potential for coastal habitats to develop. Attempts to increase biodiversity on seawalls are limited to the now restricted area of the seawall that remains submerged at low tide.

Sea level rise is emerging as a huge problem especially for low-lying coastal areas and it is inevitable that coastal defence structures need to be elevated to prevent terrestrial flooding and disruption of coastal activities. With higher seawalls becoming a common feature, it is timely to consider coastal protection together with the enhancement of coastal biodiversity. Structures on the seaward side of the seawall can be incorporated to mimic the lost intertidal area and help habitats and biodiversity to re-establish so that a return to ecosystem service provision is possible.

4.0 Increasing Biodiversity on Urban Structures

Coastal urbanization need not be equated with biodiversity elimination. While original habitats together with the intertidal and nearshore area, are very much reduced, new structures associated with the urban waterfront do provide a suitable substrate for species to colonize [23,24]. Studies include not only colonization by epibiotic or encrusting communities [25–28], but also associated pelagic communities that are attracted to the vicinity [29,30].

Jetty piles, e.g., extend the entire water column offering the range of environmental conditions present on intertidal flats [30–32]. Floating pontoons offer settlement opportunities for epibiotic species less tolerant of tidal exposure [33] and contribute to the trophic dynamics of the nearshore [29]. Increasingly more artificial structures will appear in urban waterfronts as urban sprawl intensifies and there is growing attention to their contribution to biodiversity support [34,35]. In particular, eco-engineering of such structures can help to overcome the restricted surface area and orientation that they presently offer.

The urban coast itself, although highly modified with the loss of the intertidal area and introduction of artificial structures still supports marine biodiversity and enhancing it will be beneficial for the ecosystem services provision. For example, the structural complexity throughout the water column (pontoons, piles, jetties) of marinas, common on urban waterfronts promoted biodiversity development. The artificial structures collectively with some surfaces shadowed and others more exposed to sunlight offer a range of microhabitats and opportunities to epibiotic and benthic species [33,36], which in turn attracted and supported pelagic species [37]. A higher proportion of opportunistic and stress-tolerant benthic species was noted within marinas compared to the sea beyond but community succession should restore a better balance particularly with the enhanced water column biodiversity. These all indicate that while a marina has displaced the original coastal habitat, it can effectively function as a replacement habitat provided the water quality is adequately managed.

Coastal biodiversity on seawalls and artificial structures develop best at or below the low-tide level where they are submerged for most of the time. The hard wall itself cannot accommodate any of the burrowing species commonly seen on natural intertidal shores but can support epibiotic species that attach to or encrust the wall's surface. Surveys of intertidal species colonizing different seawalls indicated a range of 26–51 species [38], many of which are known to inhabit Singapore's original shore habitats. Both, species richness and abundance. however, are severely restricted by the very limited space on seawalls due to the steep slope and augmentation of the challenging fluctuating conditions.

Increasing the wall's surface rugosity can create microhabitats that favor colonization by various species. These interventions do improve seawall biodiversity but to a limited extent because of the restricted area. Research into the design of concrete panels that can be retrofitted to seawalls to enhance biodiversity is expanding. These panels have pits [39,40] or have surfaces structured with grooves or textured with various geometric designs [41]. Such surface complexity increases surface area and microhabitat variation, which improves settlement, survival, and growth of early colonizers. The trials conducted all indicated higher settled species diversity. These complex panels tested on seawalls in Singapore [42,43] revealed greater species diversity and different community compositions independent of surface area compared to simple panels with smooth surfaces.

The upper and drier zone of the seawall, at least up to the high neap tide level presents an area that can be eco-engineered to provide conditions more conducive to supporting biodiversity. This will expand its role beyond that of just protection. By enhancing biodiversity, it can now be transformed into a living seawall. Water retaining features that can be retrofitted to existing seawalls will prevent complete drying at low tide and encourage biodiversity development. These can be small pits drilled into the wall [44] or precast concrete structures of up to 10-L capacity attached to seawalls and they supported species that are otherwise unable to colonize it [45–47]. Purpose-built cavities that retain water in seawalls also helped to increase biodiversity by supporting species absent from plain seawalls [48,49].

5.0　Coastal Biodiversity Reclamation

The intertidal area on seawalls can be increased if the wall is sloped or terraced (Fig. 4) compared to a vertical one (Fig. 5). Increasing the capacity of the seawall to retain water at ebb tide is possible if it is constructed with tidal pools incorporated on its seaward side (Fig. 6). On natural shores, tidal pools left by the receding tide support species including fish that cannot withstand exposure to air.

Fig. 4: A Sloped, Terraced Seawall Provides Greater Surface Area for Colonization by Epibiotic Species

Fig. 5: Vertical Seawall Offers Minimal Suitable Space for Epibiotic Colonization

Fig. 6: Seawall With Tidal Pools That Retain Water at Low Tide

A series of tidal pools of different sizes can be constructed along the seaward face of the seawall. They should be fully submerged by high neap tide to permit free movement of mobile species and adequate water exchange from one tidal cycle to the next. As the tide subsides, retained water in the pools provide conditions suitable for nearshore species until the next high tide allows them to leave and others to come in. These pools should be sufficiently deep to prevent heating of the trapped water during the day and they could be positioned at different heights of the seawall, in a way resembling terraced rice paddies. The seawall is thus eco-engineered to provide more of the intertidal environment that disappeared with coastal urbanization. This approach makes it possible to reclaim the lost intertidal zone.

On a more ambitious scale, some of the intertidal pools can be expanded to the magnitude of lagoons. The larger size of coastal lagoons offers more suitable conditions that make it easier to restore mangrove, seagrass, and reef habitats. Storm-water discharge with terrestrial sediment can be diverted to lagoons where mangroves can be planted and overflow to adjacent lagoons where seagrass and coral reef communities are being developed. Since mangrove and seagrass habitats are known for their sediment binding ability, water from the lagoons holding them would be clearer as it overflows to lagoons holding corals.

The floor of lagoons demarcated for mangrove regeneration can be raised to help the developing habitat, keep pace with the rate of sea level rise, taking into account sediment elevation. Lagoons that are demarcated for coral reef regeneration can be deeper as reef restoration can be promoted using artificial structures such as floating rafts [33] or platforms raised off the bottom (Fig. 7). Coral communities can develop well on the submerged parts of floating structures as they stay free from tidal exposure.

Fig. 7: Corals Growing on a Platform Raised Above the Seabed

This approach opens up all kinds of opportunities as some of the lagoons can be designed for aquaculture use or managed fisheries or to bring people closer to nature (Fig. 8). Enhancing coastal defence with nature is a response that ensures continued benefits of ecosystem services. This includes additional protection external to the seawall. Most of all, nature-enhanced coastal defence is a proactive way to address Sustainable Development Goal 14 (Life below water).

Fig. 8: Bringing People Up Close to Nature in a Biodiversity-Enhanced Urban Waterfront

6.0 Conclusion

A coastal defence and nearshore biodiversity reclamation zone entails a greater encroachment of sea space as the plan goes beyond a single sloped wall that offers only protection. The plan is for an extensive area of the seafront depending on how large the intertidal pools and lagoons can or will be. A mix of intertidal pools and lagoons all connected at high tide but retaining separate bodies of water at low tide will bring back nearshore biodiversity including habitats that provide valuable ecosystem services including being the first line of defence against storm surge and rising seas. We have lost much of the coast and sea to land reclamation. Now we should be transforming the urban waterfront into a nearshore biodiversity haven. We are not reclaiming the sea by converting land to sea but instead enhancing the sea by reclaiming coastal biodiversity.

Acknowledgments

I thank Ms. Ashley Tan for preparing the illustrations in Figs. 4–6 and 8. The picture in Fig. 7 is from the collection of the Reef Ecology Study Team. All other pictures are from the author.

References

(1) Lai S, Loke LH, Hilton MJ, Bouma TJ, Todd PA. The effects of urbanization on coastal habitats and the potential for ecological engineering: a Singapore case study. *Ocean Coast Manage.* 2015;103:78–85.

(2) Chou LM. Coastal ecosystems. In: Ng PKL, Corlett RT, Tan HTW, eds. *Singapore Biodiversity: an Encyclopedia of the Natural Environment and Sustainable Development.* Singapore: Didier Millet; 2011:64–75.

(3) Chang AL. Climate change defence vital to existence of low-lying country. *Straits Times.* 2019. https://www.straitstimes.com/singapore/sea-level-rise-poses-threat-to-spore. Accessed May 3, 2020.

(4) Chua TE, Bonga D. A functional integrated coastal management system towards achieving global sustainable objectives. In: Chua TE, Chou LM, Jacinto G, Ross SA, Bonga D, eds. *Local Contribution to Global Sustainable Development Agenda.* Quezon City, Philippines: Partnerships in Environmental Management for the Seas of East Asia and Coastal Management Center; 2018:525–552.

(5) Rosman JH, Koseff JR, Monismith SG, Grover J. A field investigation into the effects of a kelp forest (*Macrocystis pyrifera*) on coastal hydrodynamics and transport. *J Geophys Res.* 2007;112:C02016, doi:10.1029/2005JC003430.

(6) Smale DA, Burrows MT, Moore P, O'Connor N, Hawkins SJ. Threats and knowledge gaps for ecosystem services provided by kelp forests: a northeast Atlantic perspective. *Ecol Evol.* 2013;3(11):4016–4038.

(7) Spalding M, McIvor A, Tonneijck FH, Tol S, van Eijk P. *Mangroves for Coastal Defence. Guidelines for Coastal Managers & Policy Makers.* The Netherlands: Wetlands International and The Nature Conservancy; 2014:42.

(8)   Akber MA, Patwary MM, Islam MA, Rahman MR. Storm protection service of the Sundarbans mangrove forest Bangladesh. *Nat Hazard*. 2018;94:405–418.

(9)   Reguero BG, Beck MW, Agostini VN, Kramer P, Hancock B. Coral reefs for coastal protection: a new methodological approach and engineering case study in Greneda. *J Environ Manage*. 2018;210:146–161.

(10) Zhao M, Zhang H, Zhong Y, et al. The status of coral reefs and its importance for coastal protection: a case study of northeastern Hainan Island, South China Sea. *Sustain*. 2019;11:4354. doi:10.3390/su11164354.

(11) Christianen MJA, van Belzen J, Herman PMJ, et al. Low-canopy seagrass beds still provide important coastal protection services. *PLoS ONE*. 2013;8(5):e62413. doi:10.1371/journal. pone.0062413.

(12) Ondiviela B, Losada IJ, Lara JL, et al. The role of seagrasses in coastal protection in a changing climate. *Coast Eng J*. 2014;87:158–168.

(13) WWF. Coral reefs and mangroves act as natural barriers against tsunamis. World Wide Fund for Nature. 2005. https://wwf. panda.org/wwf_news/?17672/Coral-reefs-and-mangroves-act-as-natural-barriers-against-tsunamis. Accessed May 7, 2020.

(14) Kunkel CM, Hallberg RW, Oppenheimer M. Coral reefs reduce tsunami impact in model simulations. *Geophys Res Lett*. 2006;33:L23612. doi:10.1029/2006GL027892.

(15) Marios DE, Mitsch WJ. Coastal protection from tsunamis and cyclones provided by mangrove wetlands – a review. *Int J Biodivers Sci, Ecosyst Serv Manag*. 2015;11(1):71–83.

(16) Lee SK, Tan WH, Havnond S. Regeneration and colonization of mangrove on clay-filled reclaimed land in Singapore. *Hydrobiologia*. 1996;319:23–35.

(17) Chou LM, Tun KPP. Conserving reefs beside a marine landfill in Singapore. *Coral Reefs*. 2007;26:719.

(18) NEA. *Habitats in Harmony: the Story of Semakau Landfill*. 2nd ed. Singapore: National Environment Agency; 2009:118.

(19) PSD-C. Mangrove rescuers. Public Service Division – Challenge. *Government of Singapore*. 2020. https://www.psd.gov.sg/ challenge/ideas/feature/meet-the-transformers. Accessed May 4, 2020.

(20) Chung MG, Kang H, Choi S-U. Assessment of coastal ecosystem services for conservation strategies in South Korea. *PLoS ONE*. 2015;10(7):e0133856. doi:10.1371/journal.pone.0133856.

(21) Todd PA, Heery EC, Loke LHL, et al. Towards an urban marine ecology: characterizing the drivers, patterns and processes of marine ecosystems in coastal cities. *Oikos*. 2019;128:1215–1242.

(22) PEMSEA. *Sustainable Development Strategy for the Seas of East Asia (SDS-SEA)*. Quezon City, Philippines: PEMSEA; 2015.

(23) Dafforn KA, Glasby TM, Airoldi L, et al. Marine urbanization: an ecological framework for designing multifunctional artificial structures. *Front Ecol Environ*. 2015;13(2):82–90.

(24) Dyson K, Yocom K. Ecological design for urban waterfronts. *Urban Ecosyst*. 2015;18:189–208.

(25) Kay AM, Butler AJ. 'Stability' of the fouling communities on the pilings of two piers in South Australia. *Oecologia*. 1983;56:70–78.

(26) Huang ZG, Yan SK, Lin S, Zheng DQ. Biofouling communities on pier pilings in Mirs Bay. In: Morton B, ed. *The Marine Flora and Fauna of Hong Kong and Southern China III. Proceedings of the Fourth International Marine Biological Workshop. The Marine Flora and Fauna of Hong Kong and Southern China*. Hong Kong: Hong Kong University Press; 1992:529–543.

(27) Atilla N, Wetzel MA, Fleeger JW. Abundance and colonization potential of artificial hard substrate-associated meiofauna. *J Exp Mar Biol Ecol*. 2003;287:273–287.

(28) Andersson MH, Berggren M, Wilhelmsson D, Öhman MC. Epibenthic colonization of concrete and steel pilings in a cold-temperature embayment: a field experiment. *Helgol Mar Res*. 2009;63:240–260.

(29) Moreau S, Péron C, Pitt KA. Opportunistic predation by small fishes on epibiota of jetty pilings in urban waterways. *J Fish Biol*. 2008;72:205–217.

(30) Brandl SJ, Casey JM, Knowlton N, Duffy JE. Marine dock pilings foster diverse, native cryptobenthic fish assemblages across bioregions. *Ecol Evol*. 2017;7:7069–7079.

(31) Chou LM, Lim TM. A preliminary study of the coral community on artificial and natural substrates. *Malay Nat J*. 1986;39: 225–229.

(32) Ong JLJ, Tan KS. Observations on the subtidal fouling community on jetty pilings in the southern islands of Singapore.

In: Tan KS, ed. *Contributions to Marine Science.* Singapore: Tropical Marine Science Institute National University of Singapore; 2012:121–126.

(33) Toh KB, Ng CSL, Wu B, Spatial variability of epibiotic assemblages on marina pontoons in Singapore. *Urban Ecosyst.* 2017;20:183–197.

(34) Evans AJ, Firth LB, Hawkins SJ, From ocean sprawl to blue-green infrastructure – a UK perspective on an issue of global significance. *Environ Sci Policy.* 2019;91:60–69.

(35) O'Shaughnessy KA, Hawkins SJ, Evans AJ, Design catalogue for eco-engineering of coastal artificial structures: a multifunctional approach for stakeholders and end-users. *Urban Ecosyst.* 2019;23:431–443.

(36) Ng CSL, Toh KB, Toh TC, Distribution of soft bottom macrobenthic communities in tropical marinas of Singapore. *Urban Ecosyst.* 2019;22(3):443–453.

(37) Toh KB, Ng CSL, Leong WKG, Jaafar Z, Chou LM. Assemblages and diversity of fishes in Singapore's marinas. *Raffles B Zool.* 2016;S32:85–94.

(38) Lee AC, Tan KS, Sin TM. Intertidal assemblages on coastal defence structures in Singapore 1: a faunal study. *Raffles B Zool.* 2009;S22:237–254.

(39) Moschella PS, Abbiati M, Åberg P, Low-crested coastal defence structures as artificial habitats for marine life: using ecological criteria in design. *Coast Eng J.* 2005;52:1053–1071.

(40) Witt M, Sheehan E, Bearhop S, Assessing wave energy effects on biodiversity: the Wave Hub experience. *Philos T R Soc A.* 2012;370:502–509.

(41) Borsje BW, Van Wesenbeeck BK, Dekker F, How ecological engineering can serve in coastal protection. *Ecol Eng.* 2011;37:113–122.

(42) Loke LH, Jachowski NR, Bouma TJ, Ladle RJ, Todd PA. Complexity for Artificial Substrates (CASU): software for creating and visualizing habitat complexity. *PLoS ONE.* 2014;9:e87990. doi:10.1371/journal.pone.0087990.

(43) Loke LH, Todd PA. Structural complexity and component type increase intertidal biodiversity independently of area. *Ecology.* 2016;97:383–393.

(44) Firth LB, Thompson RC, Bohn K. Between a rock and a hard place: environmental and engineering considerations

when designing coastal defence structures. *Coast Eng J.* 2014;87:122–135.

(45)   Browne MA, Chapman MG. Mitigating against the loss of species by adding artificial intertidal rock pools to existing seawalls. *Mar Ecol Prog Ser.* 2014;497:119–129.

(46)   Hall AE, Herbert RJH, Britton RJ, Boyd I, George N. Shelving the coast with vertipools: retrofitting artificial rock pools on coastal structures as mitigation for coastal squeeze. *Front Mar Sci.* 2019;6:456. doi.10.3389/fmars.2019.00456.

(47)   Morris RL, Heery EC, Loke LHL, Design options, implementation issues and evaluating success of ecologically-engineered shorelines. *Oceanogr Mar Biol: An Annu Rev.* 2019;57:169–228.

(48)   Chapman MG, Blockley DJ. Engineering novel habitats on urban infrastructure to increase intertidal biodiversity. *Oecologia.* 2009;161:625–635.

(49)   Chapman MG, Underwood AJ. Evaluation of ecological engineering of "armoured" shorelines to improve their value as habitat. *J Exp Mar Biol Ecol.* 2011;400:302–313.

CHAPTER 9

Transitioning Toward Sustainable Development Through the Water–Energy–Food Nexus

Luxon Nhamo[1,2], Sylvester Mpandeli[1,3], Aidan Senzanje[4], Stanley Liphadzi[1], Dhesigen Naidoo[1], Albert T. Modi[2], Tafadzwanashe Mabhaudhi[2]

[1]Water Research Commission of South Africa, Lynnwood Manor, Pretoria, South Africa
[2]Centre for Transformative Agricultural and Food Systems, School of Agricultural, Earth and Environmental Sciences, University of KwaZulu-Natal (UKZN), Scottsville, Pietermaritzburg, South Africa
[3]School of Environmental Sciences, University of Venda, Thohoyandou, South Africa
[4]School of Engineering, University of KwaZulu-Natal (UKZN), Scottsville, Pietermaritzburg, South Africa
mabhaudhi@ukzn.ac.za; luxonn@wrc.org.za

Abstract

Sustainable development goals (SDGs) acknowledge the interlinkages between human well-being, economic prosperity, and a healthy environment, and hence, are associated with a wide range of topical issues that include the securities of water, energy, and food (WEF) resources, poverty eradication, economic development, climate change, health, among others. As SDGs are assessed through targets to be achieved by 2030 and monitored through measurable indicators, nexus planning was applied as a transformative approach to monitor and assess progress toward SDG in 2015 and 2018 using South Africa. WEF nexus-related SDGs that were evaluated include Goals 2 (zero hunger), 6 (clean water and sanitation), and 7 (affordable and clean energy). The Analytic Hierarchy Process (AHP) was used to integrate indicators for each of the reference years. Resource management and implementation of WEF-related SDGs improved by 31% (from 0.155 to 0.203) between 2015 and 2018 in South Africa but remained marginally sustainable. The assessment provided an evidence-based support framework for improved and effective management strategies to meet set SDG targets. The connections between the WEF nexus and SDGs strengthen cross-sectoral collaboration among stakeholders, unpack measures for cooperative governance and management, and supporting outcomes that arise from different cross-sectoral interventions. As food production, water provision, and energy accessibility are the major socio-economic and environmental issues currently attracting global attention, the method enhances climate change adaptation.

Keywords: water, energy, food, sustainable development

1.0   Introduction

Natural resources are being depleted worldwide and are under pressure to meet the needs of a growing population [1–3]. In 2017, over 1.06 billion people, predominantly from rural areas, had no access to safe and affordable energy, and half of these people live in sub-Saharan Africa [4]. As of 2016, some 793 million people in the world were still undernourished, and 2.4 billion had no access to improved sanitation [5]. In addition, ecosystems are degrading at an alarming rate as evidenced by a declining trend in the productivity of a fifth of the Earth's land surface covered by vegetation between 1998 and 2013

[6]. Major drivers of the stress on ecosystems include, but are not limited to, increasing demand for food due to increasing population growth and dietary transitions, accelerated economic development, urbanization, lack of transboundary cooperation, and climate variability and change [7]. Projections indicate that by 2050, the world population will increase to about 9 billion people, an increase that will see a rise in energy consumption by 80%, and a 60% increase in food demand [4,5]. These changes are occurring at a time when agriculture already consumes some 70% of available freshwater resources, a percentage that may rise due to the need to produce more crops to meet the increasing demand [8]. The gloomy outlook for world resources, exacerbated by the advent of climate change, and socio-economic inequities, resulted in the 2015 launching of the 2030 global agenda on Sustainable Development by 198 countries, members of the United Nations (UN), in an attempt to promote sustainability in resource management [9]. By then, the water–energy–food (WEF) nexus had already emerged as a polycentric approach that promotes sustainability in resource management [10].

The 17 sustainable development goals (SDGs) are a call to action by world countries, members of the UN, to promote prosperity while protecting the planet. All the 17 goals are aimed at monitoring 169 targets that collectively describe the progress of the world toward achieving a sustainable future [9]. As already alluded to, global challenges are exacerbated by stressors such as population growth, urbanization, and migration, as well as improved standards of living, which further strain already depleted resources [11]. Resource depletion is further worsened by demographic shifts, emerging growing middle class with its new lifestyle and the growing influence of climate change on the demand and supply chains of mainly WEF [12]. The SDGs were thus formulated in the context of challenges related to the insecurity of resource, climate change, and human well-being, and are designed in such a way that they recognize the interlinkages between human well-being, economic prosperity, and a healthy environment [9].

As the post-2015 focus has shifted toward implementing the SDGs and assessing progress being made toward attaining a sustainable planet by 2030, an important research challenge is to develop tools and models to monitor and evaluate implementation progress by countries and to interpret the data related to their monitoring [13,14]. Another challenge is aligning national policies and development plans with SDGs to avoid conflicts and policy incoherence [14,15]. Research and decision-making initiatives have been testing methods to effectively monitor and evaluate progress in implementing SDGs [16]. The SDGs were developed in such a way that each of the targets is assessed through one or more indicators that keep track of progress toward set targets. These indicators are the backbone of monitoring progress in the SDGs, the success of which is dependent on data availability. A major global challenge, however, is that many countries do not regularly collect data for many of the SDG indicators [17]. The unavailability of accurate and timely data from marginalized areas and people makes them "invisible," a scenario that exacerbates their vulnerability [17], and contradicts the very first commitment of the agenda 2030, which was to "leave no one behind" [18,19]. The advent of remote sensing has, however, provided solutions to some of the challenges of data unavailability by providing spatial data at large scale [20], and has been enthusiastically embraced by the international community as a way to deliver on the indicators. Several of the indicators related to landuse/cover and the environment can be monitored through remote sensing [20], which include the SDG indicators related to WEF resources.

Congruence between the WEF nexus and the SDGs has many advantages, and the WEF nexus is here proposed as a "fitting approach" for integrating and assessing SDGs implementation, which will provide pathways toward adaptation and resilience. The essence of the WEF nexus is to ensure the security of WEF resources without compromising ecosystem services. The WEF nexus enhances the coordination and sustainable planning and management of natural resources across sectors, at all levels, and spatial scales [21]. Nexus thinking is, therefore, relevant in terms of implementing and assessing the progress of the SDGs over time. Sectoral approaches in resource management will only result in inequality, exacerbate poverty, widen the gap between the rich and the poor, and increase vulnerability, yet a cross-sectoral approach as embedded in the WEF nexus could be seen as a precondition for achieving the SDGs [21,22].

As a result, the WEF nexus has become a major policy and management analytical framework for sustainable development [23,24]. There is currently a surge in global recognition of the importance of the WEF nexus in leveraging the implementation process for SDGs and subsequent monitoring and evaluation, particularly toward making informed decisions on goals, targets, and indicators [12]. As a polycentric and transformative approach, the WEF nexus supports the integration of indicators across sectors and clarifies how best resources can be allocated between competing needs, thus making the implementation of SDGs more efficient and cost-effective. The WEF nexus has become an important decision support tool that simplifies socio-ecological interactions [22,25–27].

The work reported herein assessed progress toward SDGs in South Africa between 2015 and 2018, using the WEF nexus. The focus was on the implementation progress of indicators related to Goals 2, 6, and 7. The procedure is suitable for assessing country progress in implementing the three SDGs, and is applicable at any spatial scale, depending on data availability. The procedure assessed the sustainability of resources management at the country level, providing evidence to policy and decision makers on achieving sustainability by 2030.

2.0   Methodological Framework

The study has five main components: (a) description of the WEF nexus analytical tool, (b) description of WEF nexus sustainability indicators, (c) linkages between WEF nexus indicators and related SDGs indicators, (d) results of periodic assessment and monitoring of SDGs performance and reporting, (e) analysis of the benefits of monitoring SDGs over time (Fig. 1).

We used the model developed by Nhamo et al. in 2020 [1] that defined WEF nexus sustainability indicators and calculated composite indices to establish an integrated numerical relationship

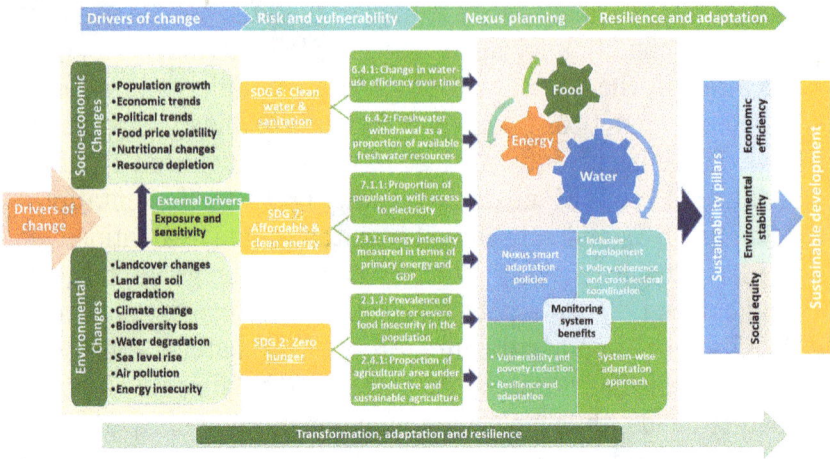

Fig. 1: Methodological Framework Linking WEF Nexus and SDG Performance and Progress

among the WEF sectors. The model provides an overview of the interlinkages among different WEF nexus indicators using the Analytical Hierarchy Process (AHP), a multicriteria decision-making (MCDM) process, to calculate composite indices for each indicator to show the relationships among the sectors, and how resources are used and managed. We identified four thematic areas involved in the interactions among WEF resources, which include (a) drivers of change, (b) risk and vulnerability, (c) nexus planning, and (d) resilience and adaptation (Fig. 1).

An assessment of country progress in implementing WEF nexus-related SDGs is achieved by evaluating resource use and management at a given time interval to achieve set SDG targets. WEF nexus sustainability indicators and the related SDGs indicators are shown in Fig. 1. Thus, progress is assessed by comparing changes over successive time intervals (three-year time interval in this study). As the WEF nexus identifies priority areas for intervention, it becomes an important decision support tool for implementing strategies that promote sustainability and ensure economic efficiency and social equity to achieve sustainability by 2030 (Fig. 1). The benefits derived from implementing the SDGs and assessing progress through the systematic and integrated approach of the WEF nexus are given at the bottom of the framework, interlinked with the other four components. The success in meeting SDG targets is assessed by the level achieved in promoting prosperity while protecting the planet [1]. Good governance is a prerequisite in resource management, as it provides the political will, and results in reducing vulnerability and poverty, enhancing adaptation to climate change, equitable resource distribution, integrated economic development, policy coherence, and intersectoral coordination. Transformation through the WEF nexus leads to adaptation, resilience, and sustainable development.

WEF resources are critical component of any economy, and they are persistently under pressure to meet the growing demands from a growing population. Increased food production will impact and depend upon water, nutrients, and energy. Thus, the knowledge on models to assess resource management, and the identification of priority interventional areas and the means to balance the economy is critical for the sustainability and resilience of food, water, and energy supply. The availability of multidisciplinary data and the development of models are prerequisites to inform policy and decision-making on agricultural intensification, water, and energy use efficiency strategies, including technology advancement.

The challenges are complex and interlinked (climate change, resource depletion, degradation and scarcity, pollution, drought, sea level rise, population increase, etc.), and they require transformative, multidisciplinary, and polycentric approaches that unpack the challenges in an integrated manner and provide holistic solution. Transformative approaches such as nexus planning, circular economy, scenario planning, and sustainable food systems are poised to provide the much-needed solutions to the pressing challenges faced by humankind.

Most African farmers rely on rainfed systems, where the duration of the rain seasons, the distribution of rains overtime varies, even more so in recent times from climate change. Under rainfed cropping systems, this would affect the cropping cycle, and consequently food availability particularly for communities that practice subsistence agriculture. Options to mitigate the consequences of water scarcity for food production include for instance water harvesting (e.g., wastewater harvesting), irrigation (e.g., drip irrigation, sprinkler systems, and partially-dry irrigation), mulching, integrated soil fertility management including enough use organic inputs, and adoption of water-use efficient crop varieties or cultivars [28]. Most of these techniques, aim especially to enhance savings on production water. Most of them are barely used in the context of Africa, due to limited scaling readiness.

A socio-economic system is composed of subsystems that are driven by other important systems such as energy, trade, and health systems [29]. The systems are linked so much so that any structural change on one food system might have originated from a change in another system, a trade-off that the WEF nexus addresses. An example of such a change is when a policy meant to promote the production of more biofuels in the energy system will obviously have a huge impact on the food system. In such cases, the WEF nexus provides policy-making with informed policy decisions that ensure sustainability and eliminating the risk of transferring challenges to other systems. Thus, the WEF nexus is meant to ensure food and nutritional security and maintaining a sustainable economic, social, and environmental base that continues meeting the needs of the present and future generations [9].

3.0   Linking WEF Nexus to SDGs 2, 6, and 7

The essence of the nexus planning is its documentation of the cross-sectoral and integrated management of resources, ensuring that any planned development in one sector is not implemented before considering the impacts (synergies, trade-offs, and implications) in the other two sectors, as well as its ability to identify different interventional priorities [1,22]. The approach is, therefore, concerned with resource sustainability and security, which are determined by factors such as availability, self-sufficiency, accessibility, and productivity, from which related indicators are defined, and applied to assess resource management and sustainability [1,30]. In each WEF sector are pillars that sustain respective indicators and contribute significantly in establishing quantitative interdependence and relationships between indicators during the comparison matrix of the AHP. Thus, WEF nexus sustainability indicators and pillars are directly linked to related SDGs and are important for evaluating the progress of SDGs implementation [1]. Both the WEF nexus and SDGs serve the same purpose of ending poverty and achieve economically and environmentally sustainable outcomes, and the former serves as an approach to spearhead the implementation of WEF nexus-linked SDGs.

SDG indicators directly linked to WEF nexus indicators (e.g., a direct measure of available water resources, or a direct measure of food security, or a direct measure of energy accessibility) are shown in Fig. 1. The focus is on indicators directly falling under the WEF nexus framework on ensuring WEF security and improving efficiency in resources management to attain sustainability. These WEF nexus values link the approach to SDGs 2, 6, and 7. The ability of the WEF nexus to convey integrated indices on resources management, and providing a glance on the status of resource sustainability over time, facilitating an assessment of progress being made toward achieving SDGs [1,21]. The study provides an assessment on the progress toward SDGs between 2015 and 2018 in South Africa as a case study. The approach indicates priority areas for intervention as a pathway to achieve sustainability.

4.0   Categorizing the Indices

WEF nexus sustainability indicators are based on ensuring resource security and are closely linked to SDGs 2, 6, and 7. These indicators are given in Fig. 1 and Table 3 [1]. Table 1 provides the classification categories for the indices to assess WEF resources performance, as well as SDGs progress. The shaded row provides the categories for the integrated WEF nexus composite index.

Table 1: Classification of WEF Nexus Indicators to Classify the Performance of Resources *Source:* Nhamo et al. [1].

Indicator	Unsustainable	Marginally Sustainable	Moderately Sustainable	Highly Sustainable
Water availability (m³/per capita)	<1,700	1,700–6,000	6,001–15,000	>15,000
Water productivity (US$/m³)	<10	10–20	21–100	>100
Food self-sufficiency (% of pop)	>30	15–29	5–14	<5
Cereal productivity (kg/ha)	<500	501–2,000	2,001–4,000	>4,000
Energy accessibility (% of pop)	<20	21–50	51–89	90–100
Energy productivity (MJ/GDP)	>9	6–9	3–5	<3
WEF integrated composite index	0–0.09	0.1–0.2	0.3–0.6	0.7–1

5.0   Pairwise Comparison Matrix and Normalization of Indices

The AHP, as pairwise comparison matrix (PCM) process, was used to determine the resource indices through relevant indicators. This was achieved by using the PCM (Equation 1) [31]. The AHP's comparison matrix compares two indicators at each time using a scale of between 1/9 and 9 as shown in Table 2 [32]. A range between 1 and 9 represents a significant relationship, and a range between 1/3 and 1/9 denotes a less important relationship (Table 2).

Table 2: Fundamental Scale for Pairwise Comparisons *Source:* Saaty [32].

Intensity of Importance	Definition	Explanation
1	Equal importance	Element *a* and *b* contribute equally to the objective
3	Moderate/weak importance of one over another	Experience and judgment slightly favor element *a* over *b*
5	Essential or strong importance	Experience and judgment strongly favor element *a* over *b*
7	Demonstrated importance	Element *a* is favored very strongly over *b*; its dominance is demonstrated in practice
9	Absolute importance	The evidence favoring element over *a* over *b* is of the highest possible order of affirmation
2, 4, 6, 8, 1/2, 1/4, 1/6, 1/8	Intermediate values between the two adjacent judgments	When compromise is needed. For example, 4 can be used for the intermediate value between 3 and 5
1/3	Moderately less important	
1/5	Strongly less important	
1/7	Very strongly less important	
1/9	Extremely less important	
Reciprocals of above nonzero	If *a* has one of the above numbers assigned to it when compared with *b*. Then *b* has the reciprocal value when compared with *a*.	

The PCM is, thus, important for determining indices, by setting priority weights to each indicator. The priority weights are set to the right of the PCM [33]. The overall significance of each indicator is then determined. The basic variable is the pairwise matrix, A, of n criteria, determined based on scaling ratios, which is of the order $(n \times n)$ [34]. A is a matrix with elements a_{ij}. The matrix is reciprocal and is expressed as:

$$a_{ij} = \frac{1}{a_{ij}} \tag{1}$$

Once the matrix is generated, it is then normalized as a matrix B, in which B is the normalized matrix of A, with elements b_{ij} and expressed as:

$$b_{ij} = \frac{a_{ij}}{\sum_{j=1}^{n} a_{ij}} \tag{2}$$

Each weight value w_i is calculated as:

$$w_i = \frac{\sum_{j=1}^{n} b_{ij}}{\sum_{i=1}^{n} \sum_{j=1}^{n} b_{ij}} i, j = 1, 2, 3, ..., n \tag{3}$$

The weighted average of the indices is considered as the integrated WEF index. The composite index indicates the overall performance of resource management. The WEF nexus integrated index portrays the relationship among the indicators in numerical form, demonstrating the scale of resources management. This index is classified as either unsustainable, marginally sustainable, moderately, or highly sustainable classes.

5.1 Calculating the Consistency of the PCM

It is important to determine the consistency ratio (CR) when using the AHP. The CR indicates whether the matrix judgments were arbitrarily selected [35]. Permissible CR values are below 0.10. When they are higher than 0.10, the PCM must be redone. The CR is calculated as [36]:

$$CR = \frac{CI}{RI} \tag{4}$$

where: CI is the consistency index, RI is the random index, the average of the resulting consistency index depending on the order of the matrix given by Saaty [32]. CI is calculated as:

$$CI = \gamma - \frac{n}{n-1} \tag{5}$$

where, γ is the principal eigenvalue, and n is the number of criteria or subcriteria in each PCM.

5.2 Overview of Resource Use in South Africa in 2015 and 2018

The WEF nexus analytical model has simplified the interpretation of the intricate relationships between the interlinked WEF resources. This quantitative evidence of the interconnectedness of the three resources provides the basis to identify priority areas for intervention that would lead to sustainability [1]. An overview of resource use in South Africa between 2015 and 2018 is given in Table 3 [37]. These statuses, together with expert advice, as well as the pillars (Table 1) are considered during the pairwise comparison [1]. The AHP, was also used to normalize the PCM values using Equations 2 and 3, and to establish the composite indices for each indicator and the integrated WEF nexus index.

Table 3: An Overview of Resource Use in South Africa in 2015 and 2018 *Source:* World Bank Indicators (2019) [47].

Indicator	Indicator Status		
	2015	2018	Units
Proportion of available freshwater resources per capita (availability)	821.3	821.4	m³
Proportion of crops/energy produced per unit of water used (water productivity)	26.2	26.2	$/m³
Proportion of population with access to electricity (accessibility)	85.5	84.4	%
Energy intensity measured in terms of primary energy and GDP (productivity)	8.7	8.7	MJ/ GDP
Prevalence of moderate/severe food insecurity in the population (self-sufficiency)	5.7	6.2	%
Proportion of sustainable agricultural production per unit area (cereal productivity)	3.5	5.6	kg/ha

6.0   Data Sources and Availability

The recognition of the importance of the WEF nexus as a decision support tool to assess the progress in implementing SDGs has gathered momentum worldwide, however, the main obstacle to achieving this has been data unavailability. Data availability is central in informing and weighting indicators during the PCM process [1]. Even where data could be available, it is normally heterogeneous in nature [38]. Data uniformity is necessary mainly for comparison purposes, particularly between countries [39]. The variations in data collection and storage bring a host of challenges, which include data disparity, mismatch, and plurality [39]. Its availability is essential for evaluating trade-offs and synergies and reduce conflicts, vital aspects of sustainable development [40]. Therefore, data availability is key for establishing indicator weights during the PCM process.

Data at regional and national levels are generally available from open source databases like FAOSTAT, AQUASTAT, and World Bank Indicators. At the national level, data can also be sourced from national statistical agents. Importantly, where data is not readily available, existing, and planned earth observation missions present reliable and long-term sources of data at all spatial scales [41,42]. For example, the Landsat Mission provides uninterrupted land and atmospheric information dating back from 1972 to date.

The success of sustainable development hinges on the availability of reliable data at all levels [43]. Publicly available data derived from remote sensing, ground stations, or models, at any spatial scale is valuable for WEF nexus assessments. Recent advances in sensor technologies and remote sensing methods to collect, analyze, and store data have facilitated the quantification, and ultimately the establishment of numerical interlinkages between the WEF sectors, and assess progress in the implementation of the SDGs [21]. For example, water-use efficiency, crop water productivity, cropped area, and landuse change detection can be mapped and calculated using satellite data [44]. The other advantage of remotely sensed data is the ability to integrate or the fusion of data that is obtained or derived at different spatial and temporal scales, or from different satellites [45].

7.0   Composite Indices for South Africa

Table 4 provides the composite indices for each indicator from the normalized PCM for 2015 and 2018 for South Africa. The integrated indices for 2015 and 2018 are 0.155 and 0.203, respectively. The two indices are then compared to assess progress and achievements toward SDG. South Africa realized significant improvement in resources management between 2015 and 2018 (an increase of 38%), however, the level of sustainability remains classified as marginally sustainable (Table 1). The target should be to have a circular shape of the centerpiece and to achieve the highest index of sustainability for each of the indicators and ensure resource security.

The indices are quantitative relationships between the indicators, providing an overview of the state of resources management in the country. The numerical relationship and the

Table 4: Integrated Composite Nexus Indices for South Africa

Indicator	Composite Indices	
	2015	2018
Water availability	0.126	0.099
Water productivity	0.128	0.221
Energy accessibility	0.141	0.079
Energy productivity	0.111	0.199
Food self-sufficiency	0.314	0.292
Cereal productivity	0.180	0.111
WEF integrated index	0.155	0.203

changes that took place in SDG implementation is best expressed using a spider graph, which illustrates the changes over time (Fig. 2). The integrated indices are calculated as the weighted average of the indicators. Both integrated indices for the respective years are very low, a situation that exposes the unsustainability in resource management, which is as a result of the sectoral approach in resources management [1]. The CRs for the pairwise matrices for 2015 and 2018 are 0.035 and 0.012, respectively, and are within the acceptable range. Any change in any one of the sectors affects the others.

7.1   Progress in Implementing SDGs in South Africa

The calculated indices (Table 4) are shown as a spider graph (Fig. 2), which provides a clear synopsis of the changes that took place in resources management in 2015 and 2018, indicating the progress in implementing SDGs in South Africa. The numerical outline shown in the spider graph highlights the interactions, interconnectedness, and interdependences of the intricate relations between the WEF sectors as seen together, providing an outlook and general progress in SDGs implementation. The further away an indicator is from the center of the axis, the greater the level of its sustainability, or the closer to the center of the axis then the closer it is to unsustainability. The irregular shape of the spider graphs shows that South Africa still must do more to achieve sustainability in resource management. The irregular shape is an indicator of an imbalanced development and mostly emanates from sectoral policies and strategies.

The evidence derived from the graphs is that South Africa focused more on food security (food self-sufficiency) in 2015, and in 2018 the country also performed well in water productivity. However, these improvements came at the expense of other resources as they contracted, giving the irregular shape (Fig. 2). A balanced and sustainable economy is achieved when the spider graph becomes circular and further out from the center. The essence of the analysis is to identify and address or enhance trade-offs or synergies, respectively.

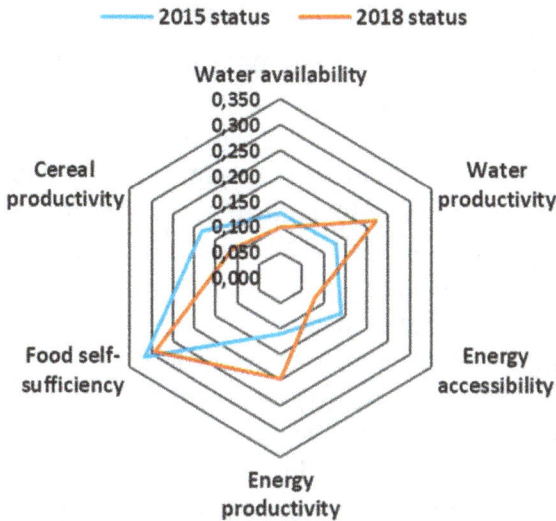

Fig. 2: A Comparison in the Progress Toward SDGs Between 2015 and 2018 in South Africa. The Irregular Shapes of Graphs Indicate an Imbalanced Resource Management

The improved water productivity index in 2018 could have been informed by the interventions that were implements during and after the severe drought of the 2015/16 rainy season [46]. The same applies with the improvement in energy productivity, which could have been as a result of interventions made after the energy crisis that took place in 2017 (also caused the reduction in energy accessibility as compared to 2015, which decreased by 1.3% between 2015 and 2018). Thus, "the good performance" in food self-sufficiency and water and energy productivity was mainly informed by prevailing challenges. The interventions were reactionary and not proactive as they were based only on solving the prevailing challenges. However, such uninformed interventions only exacerbate existing challenges. Current challenges require informed strategies from nexus planning to sustainably manage resources.

As already alluded to, a balanced and sustainable resource management requires that all indicators attain the highest index of the "best" performing indicator, without compromising other indicators, resulting in a circular shape of the spider graph. By contrast, the current sectoral approach will continue to create an imbalance in the economy and retard development, consequently slowing progress toward achieving the 2030 agenda [1]. A balanced resource management may suggest that resources are being managed holistically to achieve sustainability but can still be classified as unsustainable if the indices remain at low values. An assessment of the changes in SDGs implementation taking place over time provides evidence to research and decision makers on integrating strategies aimed at operationalizing the WEF nexus to manage resources holistically from a nexus perspective. Cross-sectoral approaches in resource management usually translate to cost and time savings compared to duplication of strategies due to sector-based interventions. Nexus planning results in improved resources use efficiencies as it informs policy and decision-making on streamlining activities, which often results in greater possibilities to manage trade-offs and synergies [22].

The study has demonstrated the WEF nexus procedure to assess progress toward SDGs, but a period of 5 years interval could be better as it would show significant changes taking place. Baseline data from national statistical agencies could be the best for this analysis if readily available.

8.0   Concluding Remarks

The WEF nexus has tracked the intricate linkages between the WEF sectors and shown the progress made toward implementing the related SDGs in South Africa. Resource management remains on the lower end of unsustainability, as evidenced by increasing poverty and hunger at the household level, water scarcity, and energy insecurity. The linkages between the WEF nexus and the related SDGs in assessing socio-ecological interactions in each context has facilitated evaluating the progress of SDGs implementation. Apart from being a decision support tool in advancing the cause of the SDGs, the WEF nexus has become a pathway to reduce poverty and achieve economically and environmentally sustainable outcomes. Both the WEF nexus and the SDGs are guided by the following common principles: (a) promotion of sustainable and efficient resource use, (b) access to resources for vulnerable population groups, and (c) maintenance and support of underlying ecosystem services. These linkages have transformed the WEF nexus into a "fitting approach" to assess SDGs implementation over time, at the same time promoting the integration of indicators across sectors and reducing the risk of sector-specific SDG actions that usually result in competition between the otherwise related sectors. The advantage of the WEF nexus approach in assessing the implementation of SDGs is its capability to analyze trade-offs and synergies between indicators, indicating priority areas for intervention, which makes it a catalyst to achieve the 2030 agenda on sustainable development. The application of the WEF nexus in assessing progress in implementing the SDGs uncovers synergies and detects destructive trade-offs among different sectors, timely exposing unexpected consequences, and pinpointing priority areas for intervention, aiming at achieving sustainability. Thus, the approach promotes integrated planning, decision-making, governance, and management of resources for simultaneous WEF security, as well as job and wealth creation in the long term. In so doing, the approach enhances cooperation and reduces conflicts among sectors, increasing resource-use efficiency in the process.

References

(1) Nhamo L, Mabhaudhi T, Mpandeli S, et al. An integrative analytical model for the water-energy-food nexus: South Africa case study. *Environ Sci Policy*. 109;2020:15–24.
(2) Pimentel D. Global warming, population growth, and natural resources for food production. *Soc Nat Resources*. 4;1991:347–363.
(3) Sherbinin Ad, Carr D, Cassels S, Jiang L. Population and environment. *Ann Rev Environ Resources*. 32;2007:345–373.
(4) IEA. *Energy Access Outlook 2017: From Poverty to Prosperity*. Paris, France: International Energy Agency (IEA); *2017*:144.
(5) FAO, IFAD, UNICEF, WFP, WHO. *The State of Food Security and Nutrition in the World 2018: Building Climate Resilience for Food Security and Nutrition*. Rome, Italy: Food and Agriculture Organisation (FAO); 2018:202.

(6)   MacDicken KG. Global forest resources assessment 2015: what, why and how? *For. Ecol. Manag.* 352;2015:3–8.

(7)   Avtar R, Tripathi S, Aggarwal AK, Kumar P. Population–urbanization–energy nexus: a review. *Resources.* 8;2019:136.

(8)   Pimentel D, Berger B, Filiberto D, et al. Water resources: agricultural and environmental issues. *BioScience.* 54;2004:909–918.

(9)   UNGA. Transforming Our World: The n.d. Agenda for Sustainable Development. Resolution adopted by the General Assembly (UNGA). New York, NY: United Nations; 2015:35.

(10)   Hoff H. *Understanding the Nexus: Background Paper for the Bonn 2011 Conference.* Stockholm, Sweden: Stockholm Environment Institute (SEI); 2011:52.

(11)   Connolly-Boutin L, Smit B. Climate change, food security, and livelihoods in sub-Saharan Africa. *Reg. Environ. Change.* 16;2016:385–399.

(12)   Yumkella K, Yillia P. Framing the water-energy nexus for the post-2015 development agenda. *Aquatic Procedia.* 5;2015:8–12.

(13)   Lim SS, Allen K, Bhutta ZA, et al. Measuring the health-related sustainable development goals in 188 countries: a baseline analysis from the global burden of disease study 2015. *Lancet.* 388;2016:1813–1850.

(14)   Maurice J. Measuring progress towards the SDGs—a new vital science. *Lancet.* 388;2016:1455–1458.

(15)   Nabyonga-Orem J. Monitoring sustainable development goal 3: how ready are the health information systems in low-income and middle-income countries? *BMJ Global Health.* 2;2017:e000433.

(16)   Janoušková S, Hák T, Moldan B. Global SDGs assessments: helping or confusing indicators? *Sustainability.* 10;2018:1540.

(17)   United Nations (UN). *The Sustainable Development Goals Report 2019.* New York, USA: United Nations; 2019:64.

(18)   Stuart E, Woodroffe J. Leaving no-one behind: can the sustainable development goals succeed where the millennium development goals lacked? *Gender Dev.* 24;2016:69–81.

(19)   Weber H. Politics of 'leaving no one behind': contesting the 2030 sustainable development goals agenda. *Globalizations.* 14;2017:399–414.

(20)   Mariathasan V, Bezuidenhoudt E, Olympio KR. Evaluation of earth observation solutions for Namibia's SDG monitoring system. *Remote Sensing.* 11;2019:1612.

(21) Mabhaudhi T, Nhamo L, Mpandeli S, et al. The water–energy–food nexus as a tool to transform rural livelihoods and well-being in Southern Africa. *Int J Environ Res Public Health.* 16(16);2019:2970.

(22) Mpandeli S, Naidoo D, Mabhaudhi T, et al. Climate change adaptation through the water-energy-food nexus in southern Africa. *Int J Environ Res Public Health.* 15;2018:2306.

(23) Albrecht TR, Crootof A, Scott CA. The water-energy-food nexus: a systematic review of methods for nexus assessment. *Environ Res Lett.* 13;2018:043002.

(24) Terrapon-Pfaff J, Ortiz W, Dienst C, Gröne MC. Energising the WEF nexus to enhance sustainable development at local level. *J Environ Manag.* 223;2018:409–416.

(25) Biggs EM, Bruce E, Boruff B, et al Sustainable development and the water–energy–food nexus: a perspective on livelihoods. *Environ Sci Policy.* 54;2015:389–397.

(26) Food and Agriculture Organisation of the United Nations (FAO). *The Water-Energy-Food Nexus: A new approach in support of food security and sustainable agriculture.* Rome, Italy: Food and Agriculture Organisation of the United Nations; 2014:28.

(27) Nhamo L, Ndlela B, Nhemachena C, et al. The water-energy-food nexus: climate risks and opportunities in Southern Africa. *Water.* 10;2018:18.

(28) Food and Agricultural Organisation of the United Nations. *Climate-smart Agriculture Sourcebook.* Rome, Italy: Food and Agricultural Organisation of the United Nations; 2013.

(29) Béné C, Oosterveer P, Lamotte L, et al. When food systems meet sustainability–current narratives and implications for actions. *World Dev.* 113;2019:116–130.

(30) Bizikova L, Roy D, Venema HD, et al. *Water-Energy-Food Nexus and Agricultural Investment: A Sustainable Development Guidebook.* Winnipeg, Canada: International Institute for Sustainable Development; 2014.

(31) Saaty RW. The analytic hierarchy process—what it is and how it is used. *Math Modell.* 9;1987:161–176.

(32) Saaty TL. A scaling method for priorities in hierarchical structures. *Journal of Mathematical Psychology.* 15;1977: 234–281.

(33) Saaty TL. Eigenvector and logarithmic least squares. *Eur J Oper Res.* 48;1990:156–160.

(34) Rao M, Sastry S, Yadar P, et al. *A Weighted Index Model for Urban Suitability Assessment—a GIS Approach*. Mumbai, India: Bombay Metropolitan Regional Development Authority; 1991.

(35) Alonso JA, Lamata MT. Consistency in the analytic hierarchy process: a new approach. *Int J Uncertain Fuzziness Knowl Based Syst*. 14;2006:445–459.

(36) Teknomo K. *Analytic Hierarchy Process (AHP) Tutorial*. Manila, Philippines: Ateneo de Manila University; 2006:1–20.

(37) World-Bank. *World Development Indicators 2017*. Washington, DC: World Bank; 2017.

(38) Zuech R, Khoshgoftaar TM, Wald R. Intrusion detection and big heterogeneous data: a survey. *J Big Data*. 2;2015:3.

(39) Liu J, Yang H, Cudennec C, et al. Challenges in operationalizing the water–energy–food nexus. *Hydrolog Sci J*. 62;2017: 1714–1720.

(40) Giampietro M. Perception and representation of the resource nexus at the interface between society and the natural environment. *Sustainability*. 10;2018:2545.

(41) Giuliani G, Chatenoux B, De Bono A, et al. Building an earth observations data cube: lessons learned from the Swiss Data Cube (SDC) on generating analysis ready data (ard). *Big Earth Data*. 1;2017:100–117.

(42) Makapela L, Newby T, Gibson L, et al. *Review of the Use of Earth Observations Remote Sensing in Water Resource Management in South Africa*. Pretoria, South Africa: Water Research Commission; 2015:153.

(43) Lawford RG. A design for a data and information service to address the knowledge needs of the Water-Energy-Food (WEF) Nexus and strategies to facilitate its implementation. *Front Environ Sci*. 7;2019:56.

(44) Nhamo L, Mabhaudhi T, Magombeyi M. Improving water sustainability and food security through increased crop water productivity in Malawi. *Water*. 8;2016:411.

(45) Huang C, Chen Y, Zhang S, Wu J. Detecting, extracting, and monitoring surface water from space using optical sensors: a review. *Rev Geophys*. 56;2018:333–360.

(46) Nhamo L, Mabhaudhi T, Modi A. Preparedness or repeated short-term relief aid? Building drought resilience through early warning in southern Africa. *Water*. SA 45;2019:75–85.

(47) World Bank Indicators. *World Bank Open Data*. 2019. https://data.worldbank.org/.

Index

A

acid rain, 264–266, 285
adequacy, electric system, 22
Adobe world headquarters,
    190–192
adsorption of hydrogen, 107–108
affordable energy for all, 13–15
    access to electricity, 15–16
    electricity, 16–19
Agence Nationale pour les
    Energies Renouvelables
    (ANER), 59
Agence pour l'Economie et la
    Maitrise de l'Energie
    (AEME), 59
Agence Sénégalaise
    d'Electrification Rurale
    (ASER), 59
agricultural communities, 277
air movement, 221
air pollutants, 269
Air Quality Standards, 281
alkaline water electrolysis
    (AWE), 93
ammonia and urea production,
    hydrogen, 98
ammonia synthesis, 98–100
anthropogenic climate change,
    30–32
analytical hierarchy process
    (AHP), 316, 320
architectural features of analyzed
    buildings, 231

construction materials and
    properties, 237–240
plan and change phases of
    house2, 234–237
plan properties and change
    phases of house1,
    231–234
area/volume (A/V) ratio of the
    buildings, 222
artificial roughness, 133
autothermal reforming (ATR), 88
AWE. *See* alkaline water
    electrolysis

B

battery charging electricity, 36
Better Buildings Initiative, 177
biodiversity, 293
    increasing on urban
        structures, 299
biofuel production pathways, 30
biofuels, 117
biogas, 103, 117
    composition, 120, 122
    low calorific value, 120–121
biological production of hydrogen,
    90–91
biomass, 2, 58
    burning, 269–270
    gasification, 89
    Senegal, 59
biophotolysis, 90
bioretention systems, 284

bisulfite, 265
brickmaking, 278
BP Energy sector definitions, 4
building automation system, 180–181
buildings, 172

C

carbon dioxide, 31–32, 117, 262–264
  emission, 38, 39
carbon payback, 29
carbon-neutral labelling for biofuels, 30
carbonaceous energy sources, 2
Carnot thermodynamics, 7
catalytic partial oxidation (CPOx), 88
CCGT. *See* combined cycle power plants
CFD. *See* computational fluid dynamics
chemical hydrides, 109
chemical industry, hydrogen, 100–101
chlorides, 274
CHP. *See* combined heat and power
circular wire, 133
climate change, 30, 262–264
climatic comfort, 219
climatic features of Diyarbakir province, 225–230
coal, 264
  and natural gas-powered electricity, 36
coastal biodiversity, 300
  reclamation, 299–305
coastal defence, 293
coastal reclamation, 294

coastal urbanization, 298, 299
codes and regulations, green retrofit, 208
combined cycle power plants (CCGT), 26
combined heat and power (CHP), 29
combustion characteristics, hydrogen, 110–111
composite indices for South Africa, 325–328
  implementing SDGs in South Africa, 326–328
computational fluid dynamics (CFD), 160–161
conduction through building, 220
consistency ratio (CR) of PCM, 322–323
construction materials and properties, 237–240
convection for heat exchange, 221
convective heat transfer, 133
cost-effectiveness, green retrofit, 207
crude oil, 270

D

dark fermentation, 91
data availability, 324
data uniformity, 324
degradation of coastal environment, 298
demand for energy, 35
demolish, 172
desulphurization, 98–99
Diyarbakir province
  annual areal precipitation in, 227

architectural features of
    analyzed buildings,
    231–240
average annual
    temperature, 226
climatic features, 225–230
general features of rural
    architecture, 224–225
map of, 223
meteorological values, 230
simulation program and
    parameters, 240–242
thermal comfort limit values,
    227–229
wind speeds and directions
    in, 229
workspace and features, 223
duct aspect ratio (W/H), 140
dust, 269–270

E

EBs. *See* existing building
ecological engineering, 293
Economic Community of
    West African States
    (ECOWAS), 65
economic gain, buildings, 175–176
economic instruments, pollution,
    283–284
ecosystems, 312
EIA. *See* Energy Information
    Administration
electricity, 15–19, 58
electricity demand in Senegal, 68
electricity generation in Senegal,
    60–68
electrification rate in Senegal,
    60, 61
electrolysers, 91–92

electrolysis, 91
Empire State Building (ESB),
    183–188
end-use energy, 2
end-user tariff (EUT), Senegal, 68
energy, 58
energy burden, 13
energy conversion, 2
energy consumption, 176, 179,
    189–190
energy convertors, 6–8
energy crisis, 58, 59
Energy Information Administration
    (EIA), 4
energy insecurity, 13
energy mix transitions, 2
energy mixes, 2, 8–12
energy performance analysis
    of house1 in summer and winter
        months, 242–247
    of house2 in summer and winter
        months, 247–254
energy performance building
    plan, 221
energy performance of buildings,
    220
energy resources, Senegal, 59–60
energy sectors, 3–5
energy sources, 2, 5–6, 8
energy supply and consumption
    in Senegal. *See* Senegal
energy system of Senegal. *See*
    Senegal
energy transitions, 39
engineering solutions, emission,
    282–283
enrichment of combustible
    mixtures with hydrogen,
    102–103
  biogas, 103
  natural gas, 103

environmental impact, buildings,
176–177
estuaries and river deltas, 280
European super grid, 25
existing building (EBs), 174
explosion behavior, hydrogen, 111

F

factories, sewage works, and
large-point sources,
277–280
Federal Poverty Line (FPL), 14
final energy, 71
financial incentives, green
retrofit, 177
financial resources, green retrofit,
207–208
firewood, Senegal, 59
flammable gases, 112
floating pontoons, 299
flow angle of attack, 139–140
forest fires, 269
fossil fuels, 68
combustion, 262
in TPES, 69
FPL. *See* Federal Poverty Line
fuel cells, 101–102
fuels, 274

G

gas explosions, 110
gasoline, 26
geothermal energy, 182
Ghana-problem, 18
GHG emissions. *See* greenhouse
gas (GHG) emissions

Glastonbury House, 192
global energy consumption, 9
global governance, 9
global life expectancy, 19
global and regional pollution, 13,
262
acid rain, snow, and lakes,
264–266
biomass burning, volcanoes,
and dust, 269–270
carbon dioxide and climate
change, 262–264
oil pollution, 270–271
plastic pollution, 271–273
stratospheric ozone depletion,
266–268
governmental policies, green
retrofit, 177
gray and blue hydrogen
autothermal reforming, 88
partial oxidation of fossil fuels
with oxygen deficient,
87–88
steam reforming, 87
thermo-electrolysis, 89
thermolysis, 89
green energy
biological production of
hydrogen, 90–91
electrolysis, 91–92
hydrogen production from
biomass, 89
greenhouse gas (GHG) emission,
6, 27, 38–39176
green retrofit, 174
codes and regulations, 208
common practices, 179
building automation system,
180–181
HVAC systems, 180

insulation system, 181
lighting systems, 179
on-site energy
    generation, 182
technological
    innovations, 182
water systems, 180
cost-effectiveness, 207
drivers, 175
    economic gain, 175–176
    environmental impact,
        176–177
    governmental policies and
        financial incentives, 177
    structural stability, 178–179
financial resources, 207–208
historic buildings, 210
operation and management, 209
safety, 210
security, 210–211
split incentives, 208
of tall buildings, 197–205
technical challenges, 208
tenants' agreement, 209
green roofs and walls, 206
gross domestic product (DGP),
    United States, 14

H

Hanwha headquarters, 195–196
heat transfer area, 133
heat transfer augmentation
    methods, 133–134
historic buildings, green
    retrofit, 210
HVAC systems, 180
Hychico-BRGM pilot project, 106
hydrocarbons, 271

hydrocracking, 98
hydroelectricity, Senegal, 60
hydrogen, 84
    advantages and disadvantages,
        84–85
    alkaline water electrolysis, 93
    flammability, 113
    green storage and utilization,
        114
        solution A, 114–116
        solution B, 116–122
    limiting oxygen concertation,
        113
    production, 86–87, 116
        gray and blue hydrogen,
            87–89
        green energy, 89–92
    proton exchange membrane,
        93–94
    safety, 110–113, 116
    solid oxide electrolysis, 94–96
    storage, 85, 104, 116
        adsorption of hydrogen,
            107–108
        chemical hydrides, 109
        hydrogen injection on natural
            gas network, 109
        metal hydrides, 108–109
        pure hydrogen, 104–107
    utilization, 97, 116
        ammonia and urea
            production, 98–100
        chemical industry, 100–101
        enrichment of combustible
            mixtures, 102–103
        fuel cells, 101–102
        refinery, 97–98
        space uses, 102
hydrogen injection on natural gas
    network, 109

hydrogen square, 85
hydrolysis, 108
hydrotreatment, 97–98

I

ignition characteristics,
    hydrogen, 110
independent power producers
    (IPP), 61
    nonexhaustive list of solar
        photovoltaic and wind
        projects, 64–65
    nonexhaustive list of thermal
        power plants, 63
industrial wastewater
    treatment, 283
infiltration-based methods, 284
International Energy Outlook
    (IEO), 33
International Poverty Line
    (IPL), 14
insulation system, 181
intertidal species, 295
ion exchange membranes, 96
IPL. *See* International Poverty
    Line

J

Joseph Vance Building, 193–195

K

Kyoto Protocol, 8

L

lagoons, 303–304
LCOE. *See* levelized cost of
    electricity
legislative and technical
    approaches, pollution,
    280–282
levelized cost of electricity
    (LCOE), 39–40
lighting systems retrofit, 179
limiting oxygen concertation
    (LOC), 113
liquid biofuels, 26
liquid waste, 279
LOC. *See* limiting oxygen
    concertation
local pollution, 273
    factories, sewage works, and
        large-point sources,
        277–280
    urban air pollution, 273–276
    urban and rural runoff, 276–277
low calorific value (LCV), 117, 119

M

major retrofit, 174
mangroves, 297
maximum experimental safe gap
    (MESG), 111
maximum explosion pressure
    (Pmax), 111
maximum rate of pressure
    rise, 111

MDGs. *See* Millennium
　　Development Goals
mechanical energy, 7
MESG. *See* maximum experimental
　　safe gap
metal hydrides, 108–109
metallurgical production from
　　smelters, 278
methane, 30, 117, 263
MIC. *See* minimum ignition current
microplastics, 272
MIE. *See* minimum ignition energy
Millennium Development Goals
　　(MDGs), 23
mini grids, 25
minimum ignition current
　　(MIC), 110
minimum ignition energy
　　(MIE), 110
minimum ignition temperature
　　(MIT), 110
minor retrofit, 174
MIT. *See* minimum ignition
　　temperature
modern bioenergy, 29–30
modern biomass–biofuels, 26
modern energy, 23–27
　　modern bioenergy, 29–30
　　nuclear energy, 27–29
　　services, 23
modern fuels and electricity, 23
Montreal Protocol, 8, 280
MSW. *See* municipal solid wastes
multicriteria decision making
　　(MCDM) process, 316
multihybrid system, 16
municipal solid wastes (MSW), 30

N

National Energy Modelling System
　　(NEMS), 32
natural gas, 103
　　vs. biogas, 119
natural habitats, 299
natural resources, 312

Natural Resources Canada
　　(NRCan), 29
nearshore biodiversity, 294–298
　　urbanization impacts on, 298
NEMS. *See* National Energy
　　Modelling System
nitrogen removal performance,
　　284
nitrous oxide, 263
nonexhaustive list of solar
　　photovoltaic and wind
　　projects, 64–65
nonexhaustive list of thermal
　　power plants, 63
nuclear energy, 20, 27, 29
nuclear fission, 23
nuclear power station, 27–28
nuclear reactor energy, 3

O

oil pollution, 270–271
on-site energy generation, 182
operating reliability, electric
　　system, 22
operation and management, green
　　retrofit, 209

"Otto" cycle, 31
ozone, 263

P

pairwise comparison matrix
(PCM), 320–322
calculating the consistency of,
322–323
resource use in South Africa in
2015 and 2018, 323
Paris Agreement, 8
partial oxidation of fossil fuels
with oxygen deficient,
87–88
PCM. *See* pairwise comparison
matrix
PEM. *See* proton exchange
membrane
pesticides, 260
petroleum products, 36
photoelectrolysis, 96
photoelectrochemical process, 96
photo-fermentation, 90
physical environmental conditions,
219
plastic pollution, 271–273
pollutants, 260, 261
pollution, 260
polycyclic aromatic hydrocarbons
(PAHs), 274
poor air quality, 273
poros media storage, hydrogen,
104
possible energy mix scenarios,
32–37
power purchasing parity (PPP), 58
pressure swing adsorption
(PSA), 87

primary energy consumption, 10
primary reformer, natural gas, 99
prime mover, 6
proton exchange membrane
(PEM), 93–94
PSA. *See* pressure swing
adsorption
pure hydrogen, storage of,
104–107
purification, 100
pyrolysis, 89

R

rainwater, 279
reasonable global mix of energies,
12–13
recycling, 286
refinery, hydrogen, 97
hydrocracking, 98
hydrotreatment, 97–98
relative gap position, 141–142
relative gap width, 141
relative roughness height (e/D),
134–137
relative roughness pitch (p/e),
137–138
relative roughness width (W/w),
142–143
relative staggered element
position, 143–144
relative staggered size, 144
reliability, of electricity supply,
21–23
renewable energy, 5, 9, 19, 20, 33
resource depletion, 313
resource use in South Africa in
2015 and 2018, 323
retrofitting tall buildings, 183–196

Rhodobacter spheroids, 90
Rhodopseudomonas palustris, 90
rib element, 133
river pollution, 278
rockwall protection, 295
roughness characterization and
        effect on fluid flow, 134
    duct aspect ratio, 140
    flow angle of attack, 139–140
    relative gap position, 141–142
    relative gap width, 141
    relative roughness height,
        134–137
    relative roughness pitch,
        137–138
    relative roughness width,
        142–143
    relative staggered element
        position, 143–144
    relative staggered size, 144
roughness geometry, 155–156
rural architecture of Diyarbakir
        province, 224–225

S

safety, green retrofit, 210
SAH. *See* solar air heater
SDGs. *See* Sustainable
        Development Goals
sea level rise, 299
sea-shores, 294
seawalls, 293
secondary reformer, natural
        gas, 99
security, green retrofit, 210–211
Senegal, 56
    airports, 74
    demography dynamics, 57

Economic Community of West
    African States, 65
electrical network, 66, 67
electrical power peak, 66, 67
electricity demand, 68
electricity generation, 60–68
end-user tariff, 68
energy consumption, 71–80
energy mix in transport
    sector, 75
energy resources, 59–60
energy supply, 68–71
energy system of, 58
fossil fuels, 69, 71
GDP comparison with other
    African countries, 57
hydroelectricity, 60
installed electrical
    capacity, 66
maritime transport, 75
oil and gas fields, 60
population estimates and
    distribution, 56
renewable energy resources, 64
sea ports, 74
solar energy potential, 59
total energy consumption
    in industrial sector, 79–80
    in residential sector, 76–78
    in transport sector, 73–76
total final energy consumption,
    71–73
wind potential, 60
Senelec, 61–63, 68
small modular reactors, 28
snow, and lakes, 264–266
socio-economic system, 318
SOE. *See* solid oxide electrolysis
soft-engineered solutions,
    pollution, 283–284

solar air heater (SAH) duct, 132
  with broken V-rib, 149–150
  with discrete V-down rib, 147
  with multigap V-rib, 150–151
  with multi V-shaped roughness,
      148–149
  with roughened surface,
      147–148
  with roughened surface having
      V-rib, 145–146
  with V-rib, 144
  V-rib geometries used in,
      157–159
  with V-rib with symmetrical
      gap, 151
  V-shaped with symmetrical
      gaps, 146
solar domestic hot water
      systems, 60
solar energy, 132
solar photovoltaic (PV) panels,
      5, 23, 182
solar-powered electricity, 17
solar radiation, 221
solar thermal, 182
solid oxide electrolysis (SOE),
      94–96
  ion exchange membranes, 96
  photoelectrolysis or
      photoelectrochemical
      process, 96
  solid polymer electrolyte, 96
solid polymer electrolyte, 96
solid waste, 283
space uses, hydrogen, 102
split incentives, green retrofit, 208
steam reforming (SR) method, 87,
      98–100
storage in engineered cavities,
      hydrogen, 104

stratospheric ozone depletion,
      266–268
structural stability, green retrofit,
      178–179
super grid, 24
supplied energy, 3
Sustainable Development Goals
      (SDGs), 12, 313–315
  indicators, 319
  implementing in South Africa,
      326–328
sustainable energy, 19–20

T

Taipei 101, 189–190
technical challenges, green
      retrofit, 208
technological innovations,
      buildings, 182
tenants' agreement, green
      retrofit, 209
thermal partial oxidation
      (TPOx), 88
thermo-electrolysis, 89
thermolysis, 89, 108
tidal oscillation, 295
total energy consumption, Senegal
  in industrial sector, 79–80
  in residential sector, 76–78
  in transport sector, 73–76
total final energy consumption,
      71–73
total primary energy supply
      (TPES), 68, 69
traditional agriculture, 58
traditional biomass fuels, 5
turbines, 7
turbomachines, 7

U

ultra-supercritical coal-fired power
plants, 26
Underground Sun Conversion, 106
United Nations (UN), 9
United States, gross domestic
product, 14
urban air pollution, 273–276
urban and rural runoff, 276–277
urban coastlines, 292
urbanization impacts on nearshore
biodiversity, 298
urban layout, 284
utilization of backup systems, 59

V

vertical landscaping, 206
volcanoes, 270
V-rib geometries
performance of, 161–165
used in SAH, 157–159
V-rib roughness geometry, 134
V-shaped roughness on heat
transfer and friction factor,
144–159

W

wall's surface rugosity, 301
water–energy–food (WEF) nexus,
313–315

categorizing the indices,
319–320
composite indices for South
Africa, 325–328
data sources and availability,
324–325
linking to SDGs, 318–319
pairwise comparison matrix.
See pairwise comparison
matrix
water pollution, 283
water systems, 180
West African Power Pool
(WAPP), 65
non exhaustive list of
projects, 65
Willis Tower, 188–189
wind and water energy, 7
wind turbines, 182
wire, 133
World Energy Projection System
Plus (WEPS+), 33
workspace and features,
Diyarbakir province, 223

Z

zero-emission, 36
zero-emission vehicles (ZEV), 36